5$\frac{a}{2}$

GAIA'S REVENGE

—GAIA'S REVENGE—

Climate Change and Humanity's Loss

P. H. LIOTTA AND ALLAN W. SHEARER

Politics and the Environment

Westport, Connecticut
London

Library of Congress Cataloging-in-Publication Data

Liotta, P. H.

    Gaia's revenge : climate change and humanity's loss / P.H. Liotta and Allan W. Shearer.

        p. cm. — (Politics and the environment, ISSN 1932–3484)

    Includes bibliographical references and index.

    ISBN 0–275–98797–3 (alk. paper)

    1. Human ecology.   2. Nature—Effect of human beings on.   3. Human beings—Effect of environment on.   4. Environmental policy.   5. Environmental degradation.   6. Geopolitics. I. Shearer, Allan W. II. Title.

    GF75.L564   2007

    304.2'5—dc22   2006027228

British Library Cataloguing in Publication Data is available.

Library of Congress Catalog Card Number: 2006027228
ISBN: 0–275–98797–3
ISSN: 1932–3484

First published in 2007

Praeger Publishers, 88 Post Road West, Westport, CT 06881
An imprint of Greenwood Publishing Group, Inc.
www.praeger.com

Printed in the United States of America

(∞)™

The paper used in this book complies with the
Permanent Paper Standard issued by the National
Information Standards Organization (Z39.48–1984).

10 9 8 7 6 5 4 3 2 1

Every reasonable effort has been made to trace the owners of copyrighted materials in this book, but in some instances this has proven impossible. The author and publisher will be glad to receive information leading to more complete acknowledgments in subsequent printings of the book and in the meantime extend their apologies for any omissions.

Copyright Acknowledgments

Sections of this work appeared in earlier versions in:

Environment and Planning B: Planning and Design—Pion Limited, London ["Approaching Scenario-Based Studies: Three Perceptions of the Future and Considerations for Landscape Planning," 32 (1, 2005): 67–87.];

Futures ["Whether the Weather: Comments on 'An Abrupt Climate Change Scenario and Its Implications for United States National Security,'" 37 (6, 2005): 445–63.];

Security Dialogue ["Through the Looking Glass: Resident Hypocrisies in National and Human Security" (37 (1, March 2005); "In Search of Meaning: Human Security and Notes toward a Definition of" [by special invitation] (35 (3, September 2004): 362–63; "Boomerang Effect: The Convergence of National and Human Security," 33 (4, December 2002): 473–88; "Converging Interests and Agendas: The Boomerang Returns," 33 (4, December 2002): 495–98.];

Parameters ["Sense and Symbolism: Europe Takes On Human Security," coauthored with Taylor Owen, 26 (3, Fall 2006); "Chaos as Strategy: Stateless Actors and State Security," 32 (2, Summer 2002): 24–32; "To Die For: National Interests and Strategic Uncertainties," 30 (2, Summer 2000): 46–57.];

Naval War College Review ["From Here to There: The Strategy and Force Planning Framework," coauthored with Richmond Lloyd, 58 (1, Spring 2005): 121–37; "Still Worth Fighting Over? A Joint Response on National Interests," coauthored with James F. Miskel, 57 (1, Winter 2004): 102–8; "The Art of Reperceiving: Scenarios and the Future," coauthored with Timothy E. Somes, 56 (4, Autumn 2003): 121–32; "Still Worth Dying For: National Interests and the Nature of Strategy" 56 (2, Spring 2003): 123–38.];

Whitehead Journal of Diplomacy and International Relations ["Why Human Security?" coauthored with Taylor Owen, 7 (1, Spring 2006): 1–18.];

World Policy Journal ["Redrawing the Map of the Future," coauthored with James F. Miskel, 21 (1, Spring 2004): 15–21.].

Presentations and lectures that influenced the development of this work were given at the following:

A meeting of the United Nations Convention to Combat Desertification, Valencia, Spain;

The Jaipur Peace Foundation, Rajasthan, India;

Office of Net Assessment, Headquarters Integrated Defence Staff, United Service Institution, New Delhi, India;

The U.S. Naval War College, Newport, Rhode Island;

The United States Military Academy, West Point, New York;

Annual meetings of the International Studies Association in Montreal, Honolulu, and San Diego;

The Faculty of Civil Defence, University of Belgrade, Serbia;

The Fifth Pan-European International Relations Conference, The Hague;

The Pell Center for International Relations and Public Policy, Salve Regina University;

The Diplomatic Academy of the Ministry of Foreign Affairs of the Russian Federation, Moscow;

The Siberian Branch of the Russian Academy of Sciences, Barnaul, Siberia;

The College of the Atlantic, Bar Harbor, Maine;

A conference jointly sponsored by the Global Environmental Change and Human Security (GECHS), United Nations Environment Programme (UNEP), International Human Development Programme—Global Environmental Change Initiative (IHDP), Center for International Climate and Environmental Research—Oslo (CICERO), and the Center for the Study of Civil War (CSCW), International Peace Research Institute, Oslo (PRIO), in Asker, Norway;

The World International Studies Committee, First Global International Studies Conference, Bilgi University, Istanbul, Turkey;

The Third European Consortium on Political Research, Corvinus University, Budapest, Hungary;

The Sixth Open Meeting of the Human Dimensions of Global Environmental Change Research Community, University of Bonn, Germany;

The United Nations Environment Programme (UNEP) Headquarters, Nairobi, Kenya;

The Fletcher School of Law and Diplomacy, Tufts University, Medford, Massachusetts;

A session on human security and urban spaces organized by the Canadian Consortium on Human Security and the Canadian Ministry of Foreign Affairs and International Trade, and presented at the Liu Institute of Global Issues, University of British Columbia, Vancouver;

The NATO Defense College, Rome, Italy.

for Gaia

Humans aren't the first species to alter the atmosphere; that distinction belongs to early bacteria, which, some two billion years ago, invented photosynthesis. But we are the first species to be in a position to understand what we are doing. Computer models of the earth's climate suggest that a critical threshold is approaching. Crossing over will be relatively easy, crossing back quite likely impossible.

Elizabeth Kolbert,
Field Notes from a Catastrophe

We are Nature,
long have we been absent,
but now we return.

Walt Whitman

Contents

Acknowledgments

P. H. Liotta wishes to thank his family and the entire staff of the Pell Center for International Relations and Public Policy for their support during the writing of this work, which was completed under especially trying conditions. He could not wish for a more loving wife and daughter or a more dedicated and professional group of coworkers.

Allan W. Shearer would like to thank the Departments of Landscape Architecture at the Harvard Graduate School of Design and at the Rutgers School of Environmental and Biological Sciences for their support. In both places he has had the very good fortune of being among colleagues who share an appreciation for interdisciplinary research. More importantly, they have generously shared their time and experience to help overcome some of the logistical challenges that such work sometimes entails. He is also grateful for his contact with Harvard's Visual & Environmental Studies Department, a place where the often opposing efforts of analysis and composition are understood as complementary endeavors.

Work on this manuscript began at Lake Teletskoye, in the autonomous Altai Republic in Siberia, and ended at Villa Aventino, Rome.

— 1 —

The Gathering Storm:
An Introduction

Everybody talks about the weather, but nobody ever does anything about it.

Mark Twain

This is a book about the future. While meaning to provoke debate about climate change and human impact, we are also interested in the necessity to consider what Herman Kahn once called "Thinking the Unthinkable." To be blunt, with the continuing failure of decision makers engaged in the daily business of security to deal with global warming effectively, we are now entering a future from which we may not be able to turn back.

Indeed, the last time in history carbon dioxide (CO_2) levels were at levels similar to today's was during the time of the mid-Pliocene "warm" period—some three and a half million years ago. As one NASA scientist jokingly retorted to the science journalist Elizabeth Kolbert, "It's true that we've had higher CO_2 levels before. But, then, of course, we also had dinosaurs."[1]

In our brief examination, nonetheless, we do not mean to repeat the work of others by providing an exhaustive review of how events have changed over geological history[2] or even more narrowly since the advent of the Industrial Revolution.[3] Nor will we focus in detail on the relationships between past societies and their responses (or lack thereof) to climatic change.[4] Although we do address mind-sets than can impede effective decisions, we will not explain why national and global policies have failed to engage, mitigate, or even effectively adapt to the vulnerabilities of global environmental shifts.[5] Finally, we will not provide a template on which to examine and consider the various arguments regarding climate change and human impact. Instead, this book is about the challenges that confront us and finding ways to envision the most effective actions that may best taken.

The place to start our discussion of climate change is the planet we inhabit. Without trying to brag, our earth is a truly remarkable site. Most obviously, its ability to support life is—so far as we know—unique in the universe. Sometimes inspired by a sense of awe and sometimes motivated by a pragmatic need, humanity has tried to understand how this is condition is possible. But, despite concerted efforts over millennia, we have failed to reveal all of the secrets. Still we try. In the 1970s James Lovelock made an intellectual leap by suggesting the possibility that the earth's lands, oceans, atmosphere, and living matter could be seen as a single organism, one that self-regulated itself to support life.[6] Called the "Gaia Hypothesis" after the ancient Greeks' Earth Goddess, it was in many respects a new way of looking at the world. And it was an idea that was as evocative as it was controversial.[7] In part, it repositioned the role of the physical scientist into that of a physiologist. Whereas before there was the studying of processes that responded to natural forces, now there would be the looking for functions that served a larger body. But also, in part, the Gaia hypothesis went beyond the testable basis and boundaries of science. In particular, early versions included the notion that the complex of systems that comprise the earth seek "optimal" conditions for life. Arguing that the earth should be viewed as a living organism was one thing (and others such as biologist Lewis Thomas has suggested thinking of the planet as a single cell), but to imply a teleological direction for nature or planetary sentience was another. Although this image captured something deep inside the common imagination, many in the scientific community saw what they considered to be New Age bunk and summarily rejected the hypothesis as a whole. Still, the potential of Lovelock's new framework was sufficiently intriguing that some pursued what was testable. And over time, different versions of the hypothesis emerged, some of which minimally stated that biota influence the larger environment, others that maximally stated biota manipulate the environment to produce the best conditions for life. Somewhere in the middle is the position that planetary homeostasis—the particular balance of conditions that regulate temperature and atmospheric composition that allows life to exist—is provided *by* the biosphere, but not *for* the biosphere.[8] That is, although life as a whole has not intentionally engineered the planet for its own benefit, life does, nonetheless, contribute to the maintenance of this extraordinary circumstance. And we would highlight that as humans are a part of the biosphere, our species contributes to this complex effort. However, we are different than other species in that we can recognize the consequences of our activities, assess their impacts, and (at our best) repair the harm and learn to act differently. It is the awareness of this distinctive capacity that obliges us to be responsible.

The date by which we might start our discussion of climate change is 1988, when government scientist James Hansen testified before the United States Senate Energy and Natural Resources Committee. His message was that CO_2 was warming the atmosphere. What marks this event as notable was not the acknowledgment that the burning of fossil fuels contributed to "the greenhouse effect." No, such a possibility had been theorized in the nineteenth century by British scientist John Tyndall, and in the 1930s at a meeting of the Royal Meteorological Society it was claimed that CO_2 was already affecting the climate.[9] Nor was this event significant because four of the hottest years ever recorded had occurred in the 1980s or that the first five months of 1988 were hotter than any in modern weather annals. Instead, what marks this event as significant was the assuredness by which Hansen made his remarks. Whereas most scientists had up to that point been very cautious in stating the effects of pollution on climate, Hansen sat before some of the nation's most influential leaders and said he was "ninety-nine percent certain that a measurable warming of the planet was the result of the build-up of CO_2 and other artificial gases in the atmosphere."[10] Although not everyone in the room or across the country was as confident in the interpretation of the data as Hansen, it was nonetheless an alarm call and many heard it. The subsequent press coverage on the issue was substantial. In 1981 only 38 percent of Americans said they remembered hearing about or reading about the greenhouse effect; but, by 1989, 79 percent had.[11]

The year 1988 is also when the Intergovernmental Panel on Climate Change (IPCC) was established. Forged from the United Nations and the World Meteorological Organization, it brings together scientific experts and government representatives. Its mission is to continually assess the state of knowledge that pertains to climate change, and twice each decade it produces an assessment report. Because of its consensus-based approach, which must satisfy a very wide range of constituencies and stakeholders, each phrase is carefully, and sometimes agonizingly, debated. And although the final text can be (and often is) faulted for being too conservative in its interpretations of the available data, it is this same conservative trait which leads *most* (an important caveat) governments to take it very seriously. As noted by Tim Flannery, "If the IPCC says something, you had better believe it—and then allow for the likelihood that things are far worse than it says they are."[12]

Between the reception of Hansen's public declaration on the forcing of global warming and the creation of an international body to foster the production and dissemination of knowledge about the state of the climate, we believe 1988 was an important year. These events demonstrated a recognition that our actions have consequences, began an effort to assess the true measure of our impacts, and, most importantly, took a step toward acting differently.

This is not to say that all was resolved. Unfortunately and all too obviously, the entire global community did not equally take up the responsibility. As mentioned previously, although Dr. Hansen was confident in his interpretation of the data, not everyone was equally convinced. It must be recognized that the science behind climate studies is both complicated (it has many parts, like an oil refinery) and complex (the relationships among the parts allow for reinforcing feedback cycles, and responses may be nonlinear). At a very abstract level of conception, the climate system can seem relatively straightforward: The earth receives energy from the sun in the form of light; some of the light is reflected back into space; and the rest is absorbed and then reradiated in the form of heat. Some of this heat is trapped by atmospheric gases, like CO_2 and water vapor. (Here we should note that without greenhouse gases, the earth would be a ball of ice and rock.) But when even a few of the specific details are considered, the system becomes much more complicated.

For starters, variations in the earth's orbit affect the intensity of the received solar radiation; the unequal distribution of land around the globe means that the northern hemisphere heats differently from the southern hemisphere; clouds both reflect solar energy before it reaches the surface and trap radiant heat. In addition to clouds, dust lofted by volcanic eruptions can block the sun's energy. The actions of people can also affect the climate (beyond the production of greenhouse gases) by changing the reflective characteristics of the earth's surface—say, by cutting down a dark forest and replacing it with lighter grain crops—or through the vapor trails of aircraft, which can create a layer of artificial clouds. Furthermore, even if the relationships between these and other fixed variables can be definitively understood, the entire system is dynamic. As heat concentrates in one area, it flows to another conveyed through wind or ocean currents. And the system goes from being considered merely complicated to complex when feedbacks, which amplify effects, are considered. For example, an increase in atmospheric CO_2 allows the air to hold more heat, which in turn allows for more water vapor, which in turn allows still more heat to be held. The combination of complicatedness and complexity lead to a situation in which there can be considerable uncertainty in the numbers, and it is arguably for this reason that many may have been somewhat reticent to accept strong conclusions about the influence humans had on the environment.

But although there is uncertainty in climate systems, there is also broad consensus among researchers that humans are, indeed, contributing to global warming.[13] The lingering perception from the mainstream media that atmospheric scientists, geographers, and others investigating the issue are at odds with one another is a situation created by what is, in most circumstances, the effort to meet the best journalistic professional standards,[14] in trying to present a "balanced" description of opposing points of view. With the case of climate change, newspapers and magazines give equal column space to those who interpret the available (and increasing) evidence as showing anthropogenic effects on the environment and those who argue that there is no obvious correlation.

Such he said-she said reporting does not provide an adequate awareness that the climate change skeptics are relatively few in number. As a result, "unbiased reporting" contains and conveys a bias.

SECURITY

To better understand impacts of climate change on human needs, we focus them through the conceptual lens of security. To most, the word *security* implies, if not exclusively means, "national security," the protection of a given nation-state from threats posed by other nation-states or belligerent nonstate actors. But we mean to use the term in a more expansive sense, one that includes not only the security of nation-states with their defining territorial boundaries and political compositions, but also of people with their environmental needs and social structures.

Although this expanded notion will be new to some, it is grounded in an increasingly growing discussion among policy makers and security scholars about just what it means to be protected from harm. The intellectual starting point for this expanded notion of security might be traced to Richard Ullman, who wrote in 1983 that, "a threat to national security is an action or sequence of events that (1) threatens drastically and over a relatively brief span of time to degrade the quality of life for inhabitants of a state, or (2) threatens significantly to narrow the range of policy choices available to government of a state or to private, nongovernmental agencies (persons, groups, corporations) within the state."[15] This concise prose written near the end of the Cold War warrants a close reading. As an unconditional point, Ullman recognizes the conventional wisdom that direct actions taken by hostile enemies have security implications. But his definition also allows the possibility that more vaguely defined, less attributable, and perhaps even unintended events might also pose a menace. The dichotomy of friend and foe is no longer a sufficient basis on which to solely plan defenses when the notion of threat as *who* can become threat as *what*. So how might the personalized enemy or unpersonalized peril affect security? In his first possibility, Ullman again both accommodates a sense of conventional wisdom and allows for an expansion of it. One might just as easily imagine a severe and quick impact on the quality of life coming from a massive bombing raid or a powerful hurricane. But in his second possibility, Ullman's text is perhaps more radical in its potential to redefine security. Here he deals not with significant and tangible impacts (such as loss of life or the destruction of property) in the present or immanent future, but with the more abstract and even hypothetical restrictions on options. To be secure is not only to be protected, but to be able to act. This interpretation adds emphasis to the saying attributed to Benjamin Franklin: "They that can give up essential liberty to obtain a little temporary safety deserve neither liberty nor safety."[16]

Ullman's thesis opens the door to many kinds of many new security concerns, ranging across the spectrum of socioeconomic needs. In considering climate change as a security issue, we are more specifically building on the discussions of environmental security and human security.[17] To many, the image of the environment as a security concern was vividly portrayed in Robert Kaplan's February 1994 essay, "The Coming Anarchy," published in the *Atlantic Monthly*.[18] In it, explosive Malthusian population growth and environmental ruin trigger violence that overwhelms the response of weakened states. Kaplan's speculation was assailed by readers on individual details, but as a whole, it resonated with the public and raised pressing questions in Washington policy circles.[19] And perhaps fiction foreshadowed fact as later that year the Department of Defense cited deforestation, soil erosion, and water pollution as contributing factors to societal decay in Haiti that ultimately led to U.S. military involvement on the island.[20] Although skeptical of the generalizations which link ecological stress to conflict, Vaclav Smil notes that "While environmental degradation does not happen in a blinding flash, it does share two important characteristics with nuclear exchange. First, its spatial reach could be truly global, and second, its social and economic effects could be highly devastating."[21]

Several relationships between the environment and security contribute to the ideas of "environmental security."[22] First, environmental degradation can be a direct threat to health. Polluted air and water can carry toxins that can cause illness or death. Degraded natural systems can contribute to conditions that are ultimately unsustainable. Second, environmental degradation can be a contributing factor to conflict. Use of limited natural resources may trigger clashes between or within states. They might also result in migration. Third, environmental degradation can be a consequence of conflict. The act of fighting a war itself may harm the environment, and in some cases, the environment might be the target of tactical operations.[23] Noncombat military activities, such as training and testing, can also cause environmental harm.

As a subtopic within environmental change, climate change has, itself, been considered as one dimension of a suite of security concerns[24] and has recently been the topic of specific attention.[25] For some, like the citizens of Tuvalu, climate change poses an existential threat. Rising sea levels that are expected to result from melting ice at the poles may cover the Pacific island nation.[26] As a means of last resort, New Zealand has offered asylum. For many, many others, climate change will affect food production and access to clean water, and increase the spread of disease such as mosquito-borne malaria.

We should also acknowledge that several arguments have been made for not including environmental concerns alongside those that have conventionally been considered to influence national security.[27] First, the discussion often confuses nonrenewable economic resources, such as iron or oil, with renewable economic externalities, such as clean air and clean water. Although history does provide examples of the acquisition of the former as a primary motivation for waging war, examples of the latter have not materialized, although such wars have been threatened. Second, environmental degradation found within a given country can most often be tied to actions taken within its own borders. That is, to restate the message on Walt Kelly's poster for the inaugural Earth Day: "We have met the enemy and he is us," so the problem does not fit the conventional understanding of national security as defense from outside threats. Third, whereas the actions of an aggressor nation can generally be understood to be a purposeful and directed application of force, the generation of CO_2 or other forms of pollution is largely considered to be a by-product of some other—often beneficial—intention. Finally, it has been argued that the "securitization" of the environment will reduce the available means that can be employed to resolve the problems to those hierarchical and bureaucratic methods of statecraft which are common to nation-states.[28] Although we see some concern over what could become an unnecessary and unhelpful conflation of ideas, we also believe the environmental concerns, generally, and climate change, specifically, are challenges to the life of individuals and the welfare of societies. Therefore, security-focused inquiries such as this one are not just appropriate; they are necessary.

POLITICO-CLIMATIC DYNAMICS

In thinking about what might be done to prevent or to mitigate the onset of climate change, there are some actions that seem obviously prudent; however, it must also be recognized that taking such steps is not necessarily easy. Anticipating variation within the current climate system and the possibility of a new ecosystem equilibrium, the U.S. National Research Council has called for "no regrets" policies to reduce vulnerabilities and increase adaptive capability.[29] These kinds of steps include the protection and conservation of natural resources such as clean water, soil fertility, and biodiversity and would contribute to what ecologist Steven Carpenter and others have called "ecosystem resilience."[30] In essence, this kind of approach is predicated on the idea that maintaining the fundamental conditions and processes that enable all life to survive allows for a broader range of future options no matter what problems might arise. For example, if a farm has healthy soil, a farmer can adapt to a shorter or drier growing season by planting a different crop; but if the soil is eroded or degraded, then switching crops may not be an option.

However, "no regrets" does not mean "no pain." In 1974, climate scientist Stephen H. Schneider advocated a comprehensive package of policies and practices he called "The Genesis Strategy."[31] More than just articulating an argument for what we would now call "no regrets" initiatives intended to protect our resource base and provide stockpiles for times of crisis, he identified political and institutional mechanisms by which these conditions might be achieved. Most of his suggestions were, by his own recognition, far from revolutionary and included ideas such as the creation of national and international organizations specifically tasked with monitoring global conditions and the structuring of tax incentives directed at resource recovery. So, if such steps would be so beneficial and could be accomplished with readily recognizable means, why have they not yet been embraced by decision makers around the world? In identifying challenges to adopting a more robust attitude toward the future, Schneider astutely notes two related concerns. First, there are trade-offs in optimizing a system for efficiency and managing a system for a diversity of options. Here, a classic (and perhaps still the most illustrative) example to offer is the application of Frederick W. Taylor's principles of industrial efficiency used by Henry Ford to produce cars. Having workers specialize in a single step within a multipart assembly process allowed them to do their jobs very proficiently, and, therefore, very cost-effectively; but, the same specialization did not allow them to develop other skills with the same proficiency, certainly none that would have allowed the same economic rewards. Thus, specialization was (and is) a double-edged sword: it made these employees relatively well paid, but it also channeled their options for the future. The economic advantages that lead individuals toward specialization and away from diversification also operate in aggregate. Schneider made a point of noting the dominance of economics as an indicator of national well-being. And on this scorecard, continued growth, which can be achieved through greater efficiencies (among other ways), is prized over potential viability. Whether it be for industrial skills or agriculture (or, for that matter, academic research), specialization has its economic advantages, advantages that are too substantial to ignore. The pressures of international trade within a context of globalizing market may further exacerbate efforts to manage for diversity, rather than optimize for efficiency.

The second challenge that Schneider raises might go far toward explaining the lack of "no regrets" actions taken in the past and the continued difficulty of taking them in the future is the political calculus that must be used to weigh current needs (and wants) against future opportunities. He asked the question, "To what extent should today's governmental leaders pursue policies that enhance the stability of the future (such as building a food reserve), recognizing that this may be accomplished only by an immediate economic sacrifice that is politically risky (such as a loss of income from storing instead of selling food)?"[32] In polemical conversations on the environment, this issue is sometimes framed as whether individuals should drive small, fuel-efficient hybrid cars rather than large six- or eight-cylinder sport utility vehicles (SUVs). This choice is nontrivial when one considers the cumulative impacts on greenhouse gas emissions of all automobiles, and therefore it is important to discuss.

However, its pretext of Western industrialized consumer purchasing power and choice can unhelpfully mask pressing questions about the current vulnerabilities of many around the world and our collective expectations for the future. For example, in the face of persistent drought and a growing population, is it acceptable to plant a crop that has been genetically engineered for yield and irrigate it with groundwater? The immediate humanitarian answer is yes. But, what if it was also known that planting the crop could negatively affect the crop's natural strains through cross-pollination of nearby fields and render them unable to subsequently reproduce? Or, what if the engineered crop requires more soil nutrients than natural strains and it will significantly deplete the soil's productiveness in, say, three growing seasons? Or, what if the groundwater source for the area is replenished only very slowly and would take 100 years of "normal" rainfall to recover the amount of water used in 5 years of irrigation? Is any one of these situations sufficient to allow the local community to go hungry? What if all three of these negative impacts will occur; is that reason enough? Is there an obviously definable, or at least defensible, threshold when future potential should not be risked? More broadly, we might ask ourselves: Do future generations have rights that ensure their well-being? And if so, what are our responsibilities toward them? More pressing, how do we balance those rights—for generations who do not yet exist—with the rights of those who are alive now?

When we filter these kinds of daunting questions through a prism that splits the spectrum of security concerns, some of the challenges in answering them come into even sharper focus. The academic subdiscipline of "critical security studies" foregrounds the predicament that increasing one's own level of security often comes at the expense of decreasing another's level of security.[33] For example, if one nation creates a large army, then neighboring nations become less secure to the degree that they are unable to repel a potential invasion. When the notion of security is expanded beyond the narrow concern of state sovereignty, there is, as argued previously, a commensurate need to think beyond perceived threats—the hazards posed by the extremes of climate variability such as high rainfall events or heat waves—by including a consideration of vulnerabilities—inherent limitations or disadvantages in the material conditions and social structures that otherwise allow individuals to subsist (if not thrive) and communities to function (if not prosper).[34] And once that intellectual maneuver is done, trade-offs across kinds of security to a given stakeholder also become apparent. As Jon Barnett notes, militaries are major emitters of greenhouse gases, and hence, in providing military security, they simultaneously contribute to environmental insecurity vis-à-vis contributing to climate change.[35] As this example demonstrates, our efforts to increase our own security from threats may not *only* decrease the security of others; they may also increase our own long-term vulnerabilities.

Similarly, steps taken to lessen long-term ecological vulnerabilities may increase the risks of short-term "traditional" security threats. An example of this
dynamic can be seen in the use and stewardship of military lands. To prepare
troops for combat and test weapons systems, armed forces need training areas
that reflect the wide variety of conditions in which they might be asked to fight.
And to maintain these fundamental resources, military lands are managed under
conservation plans. But how should training and stewardship needs be balanced?
Learning to dig a foxhole quickly can save the life of a soldier. But what if digging
holes destroys critical breeding habitat for an endangered species that (adding
another dimension of complexity to environmental and social relationships) is
protected by federal law? Similarly, learning to drive a heavy vehicle over a given
kind of terrain can contribute to the success of a mission. But what if driving on
that same terrain results in soil erosion and eventually the silting of a stream and
the degradation of a wetland?[36] These kinds of dilemmas are increasingly faced
on military bases. And given that one of the expected consequences of climate
change will be a loss of biodiversity,[37] large military installations may increasingly
be called upon to conserve species and their habitats.

In light of national, environmental, and human security, "no regrets" strategies to protect societal flexibility and maintain ecological resilience from
climate change (or other contextual changes) are rarely, if ever, truly conse-
quence-free either for one's own nation-state or for others'. Therefore, when
we do think about climate change, we should consider the full ramifications of
our potential actions. This is not to suggest that options often referred to as "no
regrets" solutions should not be pursued. Indeed, after careful deliberation and
assessment of the advantages and disadvantages, they may still offer preferred
courses of action. Nor is it to imply that we are necessarily trapped in an
us-versus-them situation like the Cold War or a zero-sum game. Rather, we do
want to make the argument that the relationships between climate change and
security are complex and that when assessing alternative options, it is necessary
to think in terms of Emma Rothschild's expanded notion of security: down
from the nation-state to individual, up to the international system, and across
to other facets of society and the environment.[38] And furthermore, in addition
to this more encompassing understanding of security, we must also extend our
thinking through multiple points in the future. That is, we must consider how
our choices made today will play out in the next day, the next year, the next
decade, and the next century because our sense of near- and far-term threats
and vulnerabilities may—and likely will—change.

SCENARIOS AS A TOOL TO STRUCTURE THOUGHT

As a way to approach the relationships between climate and security, we emphasize the use of scenarios. At a fundamental level, a scenario is a story—an ac-

count of events that has identifiable characters in a specific setting, all of which interact by following some plausible logic. But at the same time, a scenario is not just *any* story. Rather, a scenario is a way to structure thoughts of a possible future in order to (1) better understand the opportunities and challenges that might lie ahead and (2) make decisions today that are advantageous to those opportunities and robust against the challenges.

Scenarios have long been used to consider strategic security options; indeed, many of the techniques by which we frame uncertainties and relationships between elements were developed by the military or defense-oriented think tanks like the RAND Corporation—and we will discuss some of these in this book. However, the specific qualities of climate change as a security concern are distinguishable from potential force-on-force speculations. Therefore, for scenarios of a different sort of *geo*-politics, we want to emphasize two points.

First, the expanded definition of security also requires a distinction between *threats,* which are commonly discussed in security debates, and *vulnerabilities,* which are always implicit but far too infrequently addressed directly. A threat is an identifiable, often immediate, cause of harm, and one that requires an understandable response. Given these specifics, it is possible to (somewhat) precisely define countermeasures. For example, military forces have been sized and organized to defend a state against specific capabilities of potential external foes. By contrast, a vulnerability is an inherent condition that is susceptible to damage or degradation. Awareness of a vulnerability is often only through an indicator, a sign that signifies a condition, but for which cause and effect relationships that could change the condition are not well understood. Moreover, a vulnerability is often linked to other interdependent systems and related issues. As such, the recognition of a vulnerability does not always suggest an obvious or even an adequate response.

Second, when thinking about the future, we must be cognizant not only of time horizons, but also of the rate of change. Richard Falk wrote, "The *first law of ecological politics* is that there exists *an inverse relationship between the interval of time* available for adaptive change and the *likelihood* and *intensity* of violence, trauma, and coercion accompanying the process of adaptation" [his emphasis].[39] One uncertainty of climate dynamics is the speed at which the system may change. During the industrial era, we have seen a gradual increase in temperatures (mediated by pollution-related atmospheric dimming after World War II). One possibility is that such gradual change could continue, and if that happens, there is a reasonably good chance that society will be able to make adjustments: For example, farmers will plant different crops, sea walls will be heightened, and so on. This is not to say these adjustments will be cheap or easy, but the change could be managed. But is it also possible that the climate could reach an as yet unknown tipping point and flip to another (relatively) stable state over a few years. One such abrupt climate change scenario involves the slowing and shifting southward of the Gulf Stream over a period of a few years. Without the heat that it conveys, the climate of northern Europe would become like that of Siberia. Beyond long-term loss of agricultural productivity, there would be an immediate shock to the continent.

The task of envisioning the future is not easy. It requires articulating hopes and acknowledging fears. It also requires assessing capabilities and appreciating limits. Still, it is important work. As well stated by Kenneth Boulding:

Unless we at least think we know something about the future, decisions are impossible, for all decisions involve choices among images of alternative futures. This is why the study of the future is more than an intellectual curiosity; it is something that is essential to the survival of humankind itself. All our decisions are about imaginary futures and are made in the imagination, but if these imaginations are unrealistic and ill informed, decisions based on them are all too likely to be disastrous.[40]

What follows is an expansion of the ideas presented in this introduction. In chapter 2, we look at a scenario of rapid climate change prepared for the Department of Defense—Office of Net Assessment. In chapter 3, we step back and look more carefully at the concept of security, focusing on the relationships between threats and vulnerabilities. In chapters 4 and 5, we discuss the principles behind creating and using scenarios. Chapter 6 presents several methods drawn from the larger literatures of futures studies and security studies to, we think, make more focused images of the future relative to context of climate change and the needs of society. Finally, chapter 7 draws this material together and offers some directions for further thought.

NOTES

1. Extracted from Elizabeth Kolbert, *Field Notes from a Catastrophe: Man, Nature, and Climate Change* (New York: Bloomsbury, 2006), pp. 127–29. Kolbert reports that data extracted from core ice samples at the Vostokl research station in Antarctica show that at 378 parts per million, CO_2 levels now exceed levels seen in recent geological history, and that the previous high of 299 parts per million occurred some 325,000 years ago.

2. For example, Richard B. Alley, *The Two-Mile Time Machine: Ice Cores, Abrupt Climate Change, and Our Future* (Princeton, NJ: Princeton University Press, 2000).

3. For a history on the recognition of climate change and the awareness of anthropogenic causes of change, see Spencer R. Weart, *The Discovery of Global Warming* (Cambridge, MA: Harvard University Press, 2003); Gale E. Christianson, *Greenhouse: The 200-Year Story of Global Warming* (New York: Walker, 1999).

4. For example, Jared M. Diamond, *Collapse: How Societies Choose to Fail or Succeed* (New York: Viking, 2005); Brian M. Fagan, *The Long Summer: How Climate Changed Civilization* (New York: Basic Books, 2004); Brian M. Fagan, *The Little Ice Age: How Climate Made History, 1300–1850* (New York: Basic Books, 2000); Brian M. Fagan, *Floods, Famines, and Emperors: El Niño and the Fate of Civilizations* (New York: Basic Books,

1999); Stephen H. Schneider and Randi Londer, *The Co-Evolution of Climate and Life* (San Francisco: Sierra Club Books, 1984).

5. For a history of anthropogenic cause of climate change, see Tim Flannery, *The Weather Makers: How Man Is Changing the Climate and What It Means for Life on Earth* (New York: Atlantic Monthly Press, 2005).

6. James Lovelock and Sidney Epton, "The Quest for Gaia," *New Scientist* 65 (February 6, 1975): p. 935; James E. Lovelock, *Gaia: A New Look at Life on Earth* (London: Oxford University Press, 1979).

7. For a history of the evolution of the Gaia Hypothesis and its various receptions, see Jon Turney, *Lovelock and Gaia: Signs of Life* (Duxford, Cambridge: Icon Books, 2003).

8. George Ronald Williams, "Gaian and Nongaian Explanations for the Contemporary Level of Atmospheric Oxygen," in *Scientists on Gaia*—Papers Delivered at the American Geophysical Union's Annual Chapman Conference, ed. Stephen Schneider and Penelope Boston, March 1988 (Cambridge, MA, 1991), pp. 167–73.

9. Weart, pp. 2ff.

10. Philip Shabecoff, "Sharp Cut in Burning of Fossil Fuels Is Urged to Battle Shift in Climate," *New York Times,* June 24, 1988, pp. A-1, A-14.

11. Weart, p. 156.

12. Flannery, p. 246.

13. One effort to identify an empirical survey of consensus was done by Richard Posner, sampling high-impact journals (i.e., those considered influential because they have been read and cited in subsequent articles and books). See Richard A. Posner, *Catastrophe: Risk and Response* (London: Oxford University Press, 2004), p. 58.

14. For an empirical comparison of coverage in the leading newspapers of the United States, see Maxwell T. Boykoff and Jules M. Boykoff, "Balance as Bias: Global Warming and the U.S. Prestige Press," *Global Environmental Change* 14 (July 2004): pp. 125–36. For additional comments on the role of the press in presenting climate change in the United States, see Ross Gelbspan, *The Heat Is On: The High Stakes Battle over Earth's Threatened Climate* (Reading, MA: Addison-Wesley, 1997) and Ross Gelbspan, *Boiling Point: How Politicians, Big Oil and Coal, Journalists, and Activists Are Fueling the Climate Crisis—and What We Can Do to Avert Disaster* (New York: Basic Books, 2004).

15. Richard Ullman, "Redefining Security," *International Security* 8 (Summer 1983): pp. 129–53; this note p. 133.

16. Anonymous, *An Historical Review of the Constitution and Government of Pennsylvania* (London, 1759).

17. For a comprehensive discussion on the evolution of environmental security, see Simon Dalby, *Environmental Security* (Minneapolis: University of Minnesota Press, 2002).

18. Robert D. Kaplan, "The Coming Anarchy," *Atlantic Monthly* 273:2 (1994): pp. 44–76.

19. Jeremy D. Rosner, "Is Chaos America's Real Enemy? The Foreign Policy Idea Splitting Clinton's Team," *Washington Post,* August 14, 1994, pp. C1–C2.

20. Sherri Wasserman Goodman, "The Environment and National Security," Remarks to National Defense University, August 8, 1996.

21. Vaclav Smil, "China's Environment and Security: Simple Myths and Complex Realities," *SAIS Review* 17:1 (1997): pp. 107–26; this note p. 107.

22. Nina Graeger, "Environmental Security?" *Journal of Peace Research* 33:1 (1996): pp. 109–16.

23. Peter J. Stoett, "Ecocide Revisited: Two Understandings," *Environment & Security* 1:3 (1998): pp. 85–102; Richard Falk, "Environmental Warfare and Ecocide," *Bulletin of Peace Proposals* 4 (1973): pp. 1–17.

24. Neville Brown, "Climate, Ecology, and International Security," *Survival* 31:6 (1989): pp. 519–32; Rob Swart, "Security Risks of Global Environmental Changes," *Global Environmental Change* 6:3 (1996): pp. 187–92; M. J. Edwards, "Security Implications of a Worst-Case Scenario in the South-West Pacific," *Australian Geographer* 30:3 (1999): pp. 311–30.

25. Jon Barnett, "Security and Climate Change," *Global Environmental Change* 13:1 (2003): pp. 7–17.

26. Jon Barnett and W. Neil Adger, "Climate Dangers and Atoll Countries," *Climatic Change* 61:3 (2003): pp. 321–37.

27. Daniel Deudney, "Environment and Security: Muddled Thinking," *The Bulletin of Atomic Scientists* 47 (April 1991): pp. 22–28.

28. Barry Buzan, Ole Weaver, and Jaap de Wilde, "Environmental Security and Societal Security," *Working Papers* No. 10 (Copenhagen, Denmark: Center for Peace and Conflict Research, 1995).

29. National Research Council (U.S.), *Abrupt Climate Change: Inevitable Surprises* (Washington, DC: National Academy Press, 2002).

30. Steven R. Carpenter, "Ecological Futures: Building an Ecology of the Long Now," *Ecology* 83 (2002): pp. 2069–83.

31. Stephen H. Schneider, "The Specter at the Feast," *The National Observer* 23, no. 27 (July 6, 1974): p. 18. For a more complete treatment, see Stephen H. Schneider with Lynne E. Mesirow, *The Genesis Strategy: Climate and Global Survival* (New York: Plenum Press, 1976).

32. Schneider with Mesirow, p. 40.

33. Richard Wyn-Jones, *Security, Strategy, and Critical Theory* (Boulder, CO: Lynne Rienner, 1995).

34. Katrina Allen, "Vulnerability Reduction and the Community-Based Approach," in *Natural Disasters and Development in a Globalizing World,* ed. Mark Pelling (New York: Routledge 2003), pp. 170–84.

35. Jon Barnett, "Security and Climate Change," *Global Environmental Change* 13:1 (2003): pp. 7–17; this note 13.

36. For a particularly acute discussion on challenges between training operations and combat training, see David Rubenson, Jerry Aroesty, and Charles Thompson, *Two Shades of Green: Environmental Protection and Combat Training,* RAND Report R-4220-A (Santa Monica, CA: RAND Corporation, 1992); for a more general discussion of the U.S. Army's changing approach to land stewardship, see David Rubenson, Jerry Aroesty, Pamela Wyn Wicinas, Gwen Farnsworth, and Kim Ramsey, "Marching to Different Drummers: Evolution of the Army's Environmental Program," RAND MR-453-A (Santa Monica, CA: RAND Corporation, 1994).

37. Chris D. Thomas et al., "Extinction Risk from Climate Change," *Nature* 427, January 8, 2004, pp. 145–48.

38. Emma Rothschild, "What Is Security? The Quest for World Order," *Dædulus: The Journal of the American Academy of Arts and Sciences,* 124 (June 1995): pp. 5398.

39. Richard A. Falk, *This Endangered Planet: Prospects and Proposals for Human Survival* (New York: Random House, 1971), p. 353.

40. Kenneth E. Boulding, "Introduction," in *The Future: Images and Processes,* ed. Elise Boulding and Kenneth E. Boulding (Thousand Oaks, CA: Sage Publications, 1995), pp. 1–3, this quote 1.

— 2 —

Whether the Weather: An Abrupt Climate Change Scenario and Its Meaning(s) for Security

That which cannot be spoken of is what passes over into silence.

Ludwig Wittgenstein

In February 2004, the Department of Defense (DoD) released "An Abrupt Climate Change Scenario and Its Implications for United States National Security."[1] Written by Peter Schwartz and Doug Randall of consulting firm Global Business Network (GBN), the study outlined a possible future with climatic conditions similar to those 8,200 years ago and speculated on some implications related to the subsequent availability of food, water, and energy. Many of the media reactions to the text included tempestuous passages; some could even be termed blustery. Headlines included the following: "The Pentagon's Weather Nightmare,"[2] "Pentagon Report Plans for Climate Catastrophe,"[3] and "Now the Pentagon Tells Bush: Climate Change Will Destroy Us."[4] Even the usually staid *New York Times* discussed this effort of strategic foresight alongside the disaster-fantasy film *The Day After Tomorrow* [5] in its story, "The Sky Is Falling! Say Hollywood and, Yes, the Pentagon."[6] As with many topics that portend widespread doom and gloom, the press coverage itself eventually became news.[7]

Although bringing possible futures to the public's attention may be beneficial toward helping make plans that are preemptive of harmful change or, at least, that offer ways to adapt to it, the sensationalist tenor of these articles may have resulted in the scenario's relegation to the archives. As Schwartz explained in a radio interview, the project's publicity, which notably included misconceptions about its intents, may have rendered the document too politically controversial for government defense planners to engage at the present time.[8] Perhaps reflecting this possibility, the *New York Times* reported that the scenario had been neither sent up the military chain of command nor circulated to high ranking officials in the Bush administration.[9]

Whether or not the Pentagon actually postponed its own deliberations over the document, its contents are of current interest to many who are actively involved in discussions about the future. Broadly, as with any scenario-based undertaking, there are questions about how the vision of the future has been crafted and, subsequently, how it may best be used to inform a decision-making process. More narrowly, this particular scenario is also potentially significant within ongoing debates on the role of environmental factors in matters of national security. This chapter—which serves as the template for all subsequent considerations in this text—considers some strengths and weaknesses of Schwartz and Randall's work in light of these concerns.

THE CONTEXT AND THE CONTENT OF THE SCENARIO

As described in the previously mentioned press reports, the abrupt climate change scenario was initiated by Andrew Marshall, Director of DoD's Office of Net Assessment (ONA). After reading the National Academy of Sciences report *Abrupt Climate Change: Inevitable Surprises,*[10] Marshall asked Peter Schwartz and colleagues at GBN to develop a scenario for the Pentagon's consideration. Schwartz is well known among the ranks of scenario consultants and has written or cowritten several books including *Inevitable Surprises,*[11] *The Long Boom,*[12] and the often cited *The Art of the Long View.*[13] He was once a member of the highly regarded strategic planning group at Royal Dutch/Shell and has done work for organizations as diverse as the Central Intelligence Agency and the media company Dreamworks SKG. The 22-page report to ONA was dated October 2003 and became widely available to the public through the Environmental Media Service (EMS) Web site the following February. EMS is a nonprofit organization that distributes environmental information to journalists.

Schwartz and Randall give caveats throughout the report (see Appendix One), which should be emphasized with any reading of it, especially given the misperceptions that were conveyed in some of the mainstream press offerings. First, their intention was to dramatize the impact of climate change and discuss its strategic implications, not to predict when or how such change would occur. Second, the climatic conditions of the scenario were developed in consultation with several climate change experts and were based on a past sequence of events for which there is paleontological evidence. Third, although the historic record supports the position that the climatic conditions of the scenario are not implausible, the assumptions about precisely which parts of the globe are likely to be colder, drier, and windier cannot be confirmed on the basis of current climate models. Furthermore, the scientists who were consulted for this project caution that these conditions are not the most likely to occur, that the magnitude of the change may be "considerably smaller," and that the impacts of such change may be limited to a few regions. Perhaps above all, the authors note that this scenario represents a low-probability, but high-impact, future.

In brief, the scenario describes a future in which the gradual rising temperatures that have been experienced since in the middle of the twentieth century continue through 2010. Worldwide, temperatures increase by 0.5 degrees Fahrenheit (F) in the first decade of the twenty-first century and by as much as 2 degrees F in some places. Weather patterns also become more erratic, floods become more common in mountainous regions, and droughts occur in coastal and inland agricultural areas. Importantly, the increasing temperatures contribute to a positive feedback loop in the global climatic system, which accelerates the rate of warming. As a result, weather patterns become even more severe and less predictable with the passage of time. Before the decade ends, rising seas and more intense storm surges render some coastal cities such as The Hague unlivable. The Greenland ice sheet shrinks as the amount of snow that falls over it is less than the amount of glacial melt sent into the Atlantic. Overall, there is increased freshwater runoff in the high-latitude areas. By 2010, the freshening of the Atlantic from this runoff so alters the density of the seawater that the ocean's thermohaline circulation (caused by differences in the temperature and salinity of sea water) slows, and the Gulf Stream, which had brought warm water and air to northern Europe, shifts southward.

Between 2010 and 2020, climatic conditions worsen. Growing seasons are reduced by 10–25 percent, and farmers must adapt to manage different pests that come with the new conditions. Fish migrate with shifting ocean temperatures and affect commercial fishing activities, which are tied to location-specific rights. In Europe, average temperatures drop by 6 degrees F. Northwestern Europe becomes much colder, drier, and windier and comparable to present-day Siberia. Southern Europe faces intermittent cooling and rapid temperature shifts. Throughout Europe, drought and soil loss contribute to food shortages. On the other side of the Atlantic, average temperatures in North America drop by up to 5 degrees F, and windier, drier conditions reduce food production. The southern states become much drier, and coastal areas are threatened by rising ocean levels. In Asia, China faces widespread famine amid rising average temperatures and greater uncertainty about the timing and extent of the once-predictable monsoon rains. In addition, long, cold winters and hot summers stress energy and water supplies. Bangladesh suffers from persistent typhoons and rising sea levels, which make much of the nation nearly uninhabitable. East Africa suffers, in small part, from higher temperatures and drought and, in large part, by food shortages.

The authors note that anticipating the effects of the thermohaline circulation slowing in the Southern Hemisphere is (even) more uncertain because of a lack of paleoclimatic data. Thus, they offer two alternatives. One possibility is that areas below the equator also become colder and drier as the thermodynamic systems move toward a global balance. The other possibility is that the Southern Hemisphere could become much warmer and wetter since the ocean currents would no longer transport heat toward the north.

In response to these environmental changes, countries will adopt either a defensive or an offensive strategy. Nations with rich and diverse natural resources, such as the United States and Australia, will become defensive by securing their borders and taking steps toward greater self-sufficiency. The United States will turn back starving people from the Caribbean, Mexico, and South America. Its energy needs will be met, in part, with continued Middle East contacts, and, in part, through more expensive alternative energy supplies, including nuclear, renewables, and hydrogen. As part of its inward turn, the United States backs out of the 1944 treaty with Mexico, which guarantees flow from the Colorado River to the Gulf of California.

Other countries whose needs exceed their available resources adopt offensive strategies. Access to water for drinking, irrigation, and transportation will be a particular source of conflict around the world. As such, rivers and lakes may be the sites of future skirmishes. The means of waging war may also become more deadly as nuclear proliferation accelerates following the depletion of hydrocarbon fuels and the rising importance of nuclear power. Internal stability could also be weakened by the lack of or access to resources. Nations that currently have difficulty maintaining order over diverse populations will be particularly vulnerable to such conditions.

Given these circumstances and postures, the scenario outlines political and military conflicts that could take place between 2010 and 2030. In Europe, Scandinavian populations push southward and the European Union (EU) pushes them back. As climatic conditions worsen, people from northern EU nations move to southern EU nations. Russia joins the EU, relieving energy concerns, but conflicts among member nations over food and water supplies develop. By the end of the time period, the EU nears collapse, and an increasing number of Europeans move to Mediterranean African countries. In Asia, mass migration leads to internal pressures and to skirmishes along the borders of Bangladesh, India, and China. Southeast Asia suffers from persistent conflict and Japan, in response to regional instability, develops force projection capability. Between 2015 and 2020, Japan and Russia form a strategic agreement to develop energy resources, and China sends troops to protect pipelines in Kazakhstan. By 2030, there are increased tensions between Japan and China, which is now suffering from internal disorder, over Russian energy supplies. For the United States, disagreements with Canada and Mexico over water escalate, as do conflicts with European nations over fishing rights. Refugee flows from Mexico, the Caribbean, and Europe are so significant that by 2020, border control is turned over to DoD. The United States also forges an integrated security alliance with its northern and southern neighbors. The nation's oil supplies are threatened by conflicts in the Middle East, and eventually, U. S. and Chinese naval forces come into direct confrontation as the result of internal instability in Saudi Arabia.

EVENTS AND EXPLANATIONS VERSUS ACTIONS AND REASONS

Because the future has not yet occurred, it offers no facts. As such, our image of the future is based entirely on assumptions, not all of which are equal in value. Some assumptions will differ by degree, as expressed in terms of the precision or accuracy of their descriptions. Other assumptions will vary by kind. In discussing the dynamics of change, one useful distinction that can be made is between *events* and *actions.*

Events are the subject of the natural scientist who attempts to understand the interrelationships of entities which react solely to the external forces of their environments. Moreover, the effects of the external forces are consistent and allow for experimental verification through testing under scientific methods. With sufficient understanding, these methods yield *explanations,* which provide the ability to make predictions: In the same environmental conditions, the same forces applied to the same objects will produce the same results, time and time again. In general, the more an explanation can be shown to be universal, the more authoritative it is.

In contrast, actions are the subject of the historian, the novelist, and, one could argue, the national security strategist, who each attempt to understand the interrelationships of people who have choice in how to react with others and within their context. Because of choice, it cannot be said that all people will respond to the same situations the same way. Nor can it even be said that over the course of several identical (or very similar) situations, the same person will respond the same way each time. Each situation requires a decision and provides the opportunity to follow a unique course of action. Also because of choice, actions cannot be verified or predicted in the same way that events can. Instead, inferences about actions lead to *reasons* for understanding actions. Just as the scientific community has a vested interest in developing scientific methods that lead to valid explanations, those in the social sciences and humanities have developed frameworks for inferring justifiable reasons. For example, Clayton Roberts posits three kinds of logic that are frequently used by historians: the logic of the situation (how events constrain or enable certain actions), the logic of dispositional traits (how specific individuals or organizations "usually" behave), and the logic of subsequent action (how sequences of action reveal larger intentions).[14] Kenneth Burke, whose work stems from literature but has been applied to anthropological and sociological studies, has argued that to understand what someone is doing and why they are doing it requires five interdependent terms: act, scene, agent, agency, and purpose.[15] In contrast to an explanation, the more a reason can be shown to be specific to a given case, the more authoritative it is.

The differences between events-explanations and actions-reasons are in some ways subtle but offer an important distinction for the interpretation of assumptions about the future. In earlier writings, Schwartz discusses what Pierre Wack, another corporate strategist from Royal Dutch/Shell, calls *predetermined elements:* "those events that have already occurred (or that almost certainly will occur) but whose consequences have not yet unfolded."[16] The now-classic exemplar of this notion is of a storm at the headwaters of a river, which will produce a future flood downstream. The predictive power of a scientifically based explanation allows a future event to be similarly treated within the scenario as a kind of pre-determined element because when the event is initiated, it will be followed with inevitable consequences. In the case of this scenario on abrupt climate change, it is a matter of scientific explanation that *if* the Gulf Stream shifts to the south, then northern Europe will face lower temperatures and increased drought and, commensurately, reduced plant (notably, crop) productivity.

Formal approaches for understanding reasons for action can offer some basis for a rational expectation of future occurrences, but they do not offer the same level of certitude that explanations of events do. To be sure, history does provide some examples of actions that triggered a sequence of automatic consequences, and these might be considered predetermined scenario elements. For instance, historian A. J. P. Taylor argues that the highly engineered and fixed railroad schedules of the early twentieth century and the lack of a formal decision-step between mobilization and combat operations significantly contributed to the start of World War I. As a result of this plan, once the call to arms was sounded, war was inevitable.[17] However, such "event-like" actions or action chains are very rare. To resort to an obvious example, as president, staunch anti-communist Nixon went to China and expanded relations, and his choice to do so was against many expectations. Hence, for the person interested in the consequences of climate change, the framework that one might use to assert and to assess an explanation that a southward shift in the Gulf Stream will lower temperatures in northern Europe, make drought more likely, and crops more difficult to grow is not the same as the framework one might assert and assess the reasons regarding how the people or the states of northern Europe will react to cold, thirst, and hunger.

CLIMATE CHANGE

Scenarios dealing with environmental change can include both events and actions, and the particular issue of climate change is no exception. As revealed by paleoclimatic evidence, the earth has gone through several warming and cooling cycles, which have progressed independent of human activities. Factors that may contribute to these cycles include variations of the earth's axis of rotation or its solar orbit and the partial blockage of sunlight by volcanic ash erupted into the atmosphere. However, some scientists point to evidence that over the twentieth century, global warming has been at least partially influenced by greenhouse gas emissions associated with the use of fossil fuels and through land use changes related to deforestation, desertification, and the construction of urban environments.[18] There is stronger evidence that since 1950, human activity has been the dominant influence on climate change.[19]

Teasing apart the relative influence of natural events and human actions on climate change is further complicated by the complexity of the global climatic system and several dimensions of uncertainty.[20] Historic climate data are sparse, and the data that do exist represent too brief a time span to encompass the full range of climatic conditions that have enveloped the earth over its history. There are also uncertainties related to the spatial and temporal resolutions of mathematical models, and there are uncertainties associated with the parameterization of variables used in the models. Data and model limitations can be partially compensated through the use of expert judgment, but there are challenges associated with resolving varying opinions.[21] Indeed, it is the case that there have been conflicting assessments of the net impacts of climate change on ecosystems.[22] Findings such as these reflect that the explanations offered by experts on this topic are provisional and constantly evolving. There have been calls for further research to address these uncertainties,[23] and such efforts will undoubtedly improve the understanding of climate change. But in the meantime, it is and will be the case that the explanations offered for climate change are relatively weaker than explanations offered for other phenomena.

This lack of a consensus about the science of climate change has had implications on policy debate regarding how actions might be or should be tempered to address future societal needs. One study of expert judgments found almost no agreement about the effect of climate change on policy-relevant factors such as changes in precipitation over land and various forms of interannual variability.[24] Even if there were such agreement, it is not clear that findings would or could be readily applied by decision makers. Another study found that the language used by experts to describe and assess the uncertainties of climate change is different from that of lay readers. As a result, nonexperts, including most members of the public and many policy makers, may underestimate the probability of high-magnitude possible outcomes.[25] It has also been argued that because the knowledge of climate change is so incomplete, attempts to assign probabilities are flawed and may lead to poor decisions.[26] Perhaps unsurprisingly, the widespread concern over climate change coupled with the difficulty of applying research findings to action has led to opposing policy stances. On one side, some have argued that despite the uncertainty in the science and the difficulty of applying admittedly partial knowledge, the extensive (and growing) literature on the possible consequences of different and prolonged weather patterns warrant some kind of global regulatory action that would lower the likelihood of society-induced or society-assisted climate change.[27] On the other side, there are calls by others, including U.S. President George W. Bush, to be very cautious with government mandates until additional research can provide a foundation for scientifically informed action.[28]

The empirical, methodological, and practical limits of climate change knowledge present difficulties for the scenarioist who is trying to evoke an image of a future that will be salient to decision makers. If a specific model of climate change is used—say, one based on some set of human activities as the driving force—then the scenario as a whole may be dismissed as irrelevant by those who disagree with that model. Offering multiple models of change that lead to the same future may simply confuse the reader or prompt a conclusion that competing models signify a lack of any clear understanding of the system. Again, the scenario as a whole might be relegated to the nearest recycling bin. However, regardless of the mechanisms that could lead to abrupt climate change, it is still the case that the consequences of such a change would be significant. As such, they (still) warrant consideration so that society might be better prepared to meet the challenges of that tomorrow.

As noted previously, one of the caveats for the report is that Schwartz and Randall do not attempt to describe how climate change will occur; instead, they simply speculate that the globe continues to gradually warm until the abrupt transformation. By leaving the reader to speculate why the warming trend continues, the authors sidestep the potential problem that comes with problematic or ambiguous models of change. Those familiar with the sometimes caustic debate on climate change in the United States might warily suspect that by not explicitly citing human action as a significant factor of climate change, the authors were playing their tune for global warming skeptics in the George W. Bush administration. But separate from any partisan distrust and above political cynicism, leaving the details of the warming trend to the reader may serve a pragmatic purpose by helping to direct the reader's attention to the possible consequences of change and thereby get on with the challenging task of grasping how society may be affected.

There are, however, two potential concerns that come with such an unelaborated leap into the future. First, just as with the perception of an incorrect or ambiguous explanation for change, an incomplete explanation may undermine the scenario's acceptance and thereby its usefulness. In the case of this scenario, this concern is arguably mitigated by the record of the historic conditions on which the scenario is patterned. The 2002 National Research Council study on abrupt climate change, which is said to have precipitated the

ONA's interest, noted: "Because so many Holocene climate records are available and the cause of the event [8,200 years ago] is rather clear, it provides an opportunity for an especially well-documented test case of model sensitivity. The event is also important because it punctuated a time when temperatures were similar to or even slightly above recent levels, demonstrating that warmth is no guarantee of climate stability."[29] Hence, although not notably emphasized in the report, knowledge of the climatic conditions 8,200 years ago is relatively good and is (at least partially) analogous to our own time. And though this knowledge is incomplete, it may provide the best starting point to begin to speculate on future conditions following abrupt climate change. That is, particular to climate change, this scenario-as-event can be reasonably well supported, even if a consensus explanation cannot be had.

The second concern about focusing on an end rather than a means may be more problematic in that it could limit efforts to grasp the political and, thereby, security implications of this future. What *if* actions as well as events contribute to rapid climate change? Suppose that just before the Gulf Stream shifts southward and plunges northern Europe into a Siberia-like cold, it is demonstrated and widely accepted that (1) rising global temperatures have predominantly been caused by greenhouse gas emissions, most of these were produced by the industrialized world, and the single greatest contributor was the United States; (2) temperature change has led to the melting of Arctic ice; and (3) the melting of the ice triggered the collapse of the thermohaline circulation.

Would the United States feel a sense of responsibility to aid and assist nations who were not as responsible for contributing to this new and difficult world? And if so, would a defensive national (environmental) security be possible? Or, would a defense strategy be thrust upon the United States if other nations held America as largely responsible? And, would such a perception negatively influence the United States' ability to lead global initiatives writ large? Would it lead to a political, social, and economic separation of the United States and the European Union? In many aspects of human relations, past actions influence current thoughts, beliefs, and feelings. This situation will likely be true in the future. As such, it can be important to factor what might be called "a forward look of history" into scenarios.

NATIONAL SECURITY AND THE ENVIRONMENT

Whereas debate over the science of climate change may be unsettled by problems of uncertainty and the analytical mixing of events and actions, discussions about the link between national security and the environment are clouded by questions over definitions and conceptual relationships. Throughout the twentieth century, national security was largely understood as the defense of sovereign nation-states through military means. During the Cold

War, the goal of containing Communism served to further structure security theory and practice. But with the collapse of the Soviet Union, threats that had existed since the end of World War II faded, and there were calls to expand the definition of *security* to include a wider range of concerns, including the environment.[30] In broad terms, this expansion initiated what continues to be a reevaluation of what constitutes a "threat" and an increasing recognition that ecological concerns play an important role in issues relating to health, economics, and political stability.[31] More specifically, it has been argued that more holistic definitions of security would better allow for analysis of transboundary issues, which are larger than the state, and personal or regional issues, which are smaller than the state.[32] Also, the inclusion of the environment within an understanding of national security has been seen as beneficial because it raises the awareness of ecological problems to that of "high politics" and, thus, to a more prominent position on national agendas. The list of environmental topics that have been suggested for inclusion as security issues is extensive and includes— in part—acid rain, fossil fuels, natural disasters, nuclear waste, oil crises, ozone depletion, rising sea levels, and soil degradation.[33] In addition to Schwartz and Randall's scenario, the intersection of climate change and a broad understanding of security has been recently discussed by Jon Barnett.[34]

Expanded understandings of national security have also found their way into some policy documents. Possibly the first, and certainly the earliest high profile, such mention occurred in 1991 when President George H. W. Bush stated as part of his National Security Strategy that "we must manage the Earth's natural resources in ways that protect the potential for growth and opportunity for present and future generations."[35] In 1994 President Clinton asserted more emphatically that "not all security risks are military in nature. Transnational phenomena such as terrorism, narcotics trafficking, environmental degradation, rapid population growth, and refugee flows also have security implications."[36] And beyond both theoretical abstractions and political rhetoric, in 1996 the environmental issues of deforestation, soil erosion, and water pollution were recognized by then–Deputy Under Secretary of Defense (Environmental Security) Sherri Wasserman Goodman as materially contributing to societal decay in Haiti, which ultimately led to intervention by U.S. forces there in 1994.[37]

The interest in expanding the definition of national security to include environmental factors is not, however, universal, and it has been counterargued that mixing of environmental concerns with security concerns is an unnecessary and unhelpful conflation. In part, placing ecological issues into national security discussions often confuses nonrenewable economic resources, such as iron or oil, with renewable economic externalities, such as clean air and clean water.[38] Although history does provide examples of the acquisition of the former as a primary motivation for waging war, examples of the latter have only been threatened and have not, in fact, materialized. As a result, whereas the actions of an aggressor nation can generally be understood to be a purposeful and directed application of force, the generation of pollution is largely considered to be an unfortunate but not intentionally malicious result from the production of wanted goods and services. Moreover, environmental degradation found within a given country can most often be tied to actions taken within its own borders.

That is, the enemy is us, and as such the problem does not fit the conventional understanding of national security as defense from outside threats. And finally, it has been argued that the "securitization" of the environment may reduce the available means that can be employed to resolve the problems to those hierarchical and bureaucratic methods of statecraft that are common to nation-states.[39] The debate on this topic will continue.

Schwartz and Randall walk the reader through a relationship between environmental conditions and national security in three steps. In the first step, they introduce the ongoing academic debate on the role of resource constraints in interstate conflicts and refer to Peter Gleick's outline of three national security challenges posed by abrupt climate change: (1) food shortages due to decreases in agricultural production; (2) decreased availability and quality of fresh water due to flooding and droughts; and (3) disrupted access to strategic minerals due to ice and storms.

In the second step, they connect the environment to national security through the notion of *carrying capacity,* which Schwartz and Randall define as "the ability for the Earth and its natural ecosystems including social, economic, and cultural systems to support the finite number of people on the planet."[40] The authors note it has been asserted by some that given the fact that 815 million people worldwide do not receive sufficient sustenance, we are already living above carrying capacity. An abrupt climate change, such as the kind described, would result in a (further) deficit and result in deaths due to starvation and war over food, water, and energy. Concerns over carrying capacity are not new to futures studies. The most well known of these, even today, is the Club of Rome project, which was published the early 1970s.[41] Predicated on concerns of population growth and finite resources, the work has been highly influential[42] but has also been criticized for overgeneralizing some of its assumptions.[43] Similar criticisms could also be applied to Schwartz and Randall's scenario; however, it should be accepted—even if begrudgingly—that it would be extraordinarily difficult to create any global-scale scenario that would not suffer from some such limitations. As George E. P. Box wrote, "All models are wrong, but some are useful."[44] The pragmatic question that follows is this: How might specific generalizations affect the interpretation of the results and the use of the conclusions?

The final step to complete the path from environmental change to national security treads across some recent work by anthropologist Steven LeBlanc. Before discussing this aspect of the report, a clarification is needed: Schwartz and

Randall refer to LeBlanc's book, *Carrying Capacity*; however, no such work exists[45] and they instead (apparently) refer to *Constant Battles: The Myth of the Peaceful, Noble Savage,*[46] which was highlighted in the book club section of the GBN Web site. The book's thesis is that the notion that people once lived in harmony and balance with their surroundings has no basis in fact. Instead, humans have persistently faced limited resources and ecological shortages have been the fundamental cause of warfare.

LeBlanc defines carrying capacity more generally as "the limitations of the local environment to support a population of animals, including people."[47] His definition also comes with several qualifying statements, including the fact the carrying capacity of a given area is not fixed but can be altered by new crops, new technology, or new lifestyles. Although there are examples of rapid increases in carrying capacity—such as the introduction of the potato to Europe—most expansions are incremental and occur over long periods of time. Climate change is also cited as a factor that can significantly affect carrying capacity. Drawing on archaeological evidence, LeBlanc presents a case that peoples from around the world have continually exceeded the carrying capacity of their environments, and when this situation occurs, conflict within a group or between groups has commonly resulted.

Capitalizing on LeBlanc's general proposal, Schwartz and Randall claim, "Every time there is a choice between starving and raiding, humans raid."[48] However, LeBlanc's argument is not so simple, nor does it go quite so far. LeBlanc distinguishes a spectrum of societies based on the complexity of socio-economic-political organization. From the simpler to the more complex, the kinds of societies are tribes, bands, chiefdoms, and states. Each kind of society is characterized by specific relationships with the environment and toward warfare. For example, and in part, with simpler societies, warfare is common, and a larger percentage of the population is involved in fighting. The act of warfighting is understood as personal and final, and as such, prisoners are not taken. The result of conflict is usually the dispersal or the death of the losing side. As societies become more complex, warfare becomes less common but can be much more intense. The conduct of war is handled by an increasingly smaller percentage of the population who are professionalized through the creation of standing armies. Soldiers on both sides of a conflict are "just doing their jobs" and so, though warfare is still emotional, it does not carry the same ultimate expectation to kill or be killed. The typical result of such warfare is that the losing population is incorporated into the winning polity.

An important aspect of complex societies is the influence of the state's central authority over the populace and how this power may be factored into the political calculus of state-level decision making. LeBlanc notes that Marie Antoinette's purported quip, "Let them eat cake," was a euphemism for "Let the peasants starve."[49] More generally, LeBlanc writes:

High levels of conflict are observed among most chiefdoms, but warfare among states actually is less intense and less and less demographically relevant than for foragers and tribes. There are a number of explanations for this phenomenon, but one is simply that people living in states will starve before they fight because the government won't allow them to fight. Moreover, such great crises as famines and other disasters knock the state's population back down to a sustainable level, so the problem is periodically—though only temporarily—solved.[50]

The possibility that a state would willingly allow its own to suffer and die, while certainly unnerving, is notable because it removes any rationalization that warfare is environmentally determined, even if it is, as LeBlanc argues, ecologically motivated; that is, going to war is an action, not an event. Furthermore, this observation introduces the consideration that state-level decisions can be based on a variety of competing concerns. And because there are trade-offs, additional details are needed to assess the rationale for any actions that may be taken. At a minimum, there must be some discussion of the relative importance of competing interests. There may also be the need to establish that any identified goals could viably be achieved. Similarly, a basis for assessing such trade-offs is also needed to understand conflict within a state. LeBlanc's example of eighteenth-century France also serves to show that a state's decision to not expand its resource base (be it through warfare or other means) may have internal consequences. There are limits to the degree that a state can ignore the plight of its people; the same French peasants that were starving opted to pursue a course of revolution. But what is the political calculus of the people? What threshold must be crossed before a populace decides to not only take to the streets in protest, but to overthrow a government and thereby potentially create a security vacuum that puts personal property and life at risk?

Relative to Schwartz and Randall's scenario, one must ask if the actions that follow the climate change events were sufficiently reasoned. The answer is "not necessarily." The difficulty in making sense of the actions in this scenario is perhaps most evident in what may be the most specific decision mentioned in the text: The United States' backing out of the 1944 treaty with Mexico that guarantees water flow from the Colorado River. Given the "defensive" strategic posture of the United States, maintaining all available fresh water seems reasonable. However, one might question whether this decision makes sense in the larger context of environmental and political concerns. One of the primary reasons Mexico has advocated for persistent flow of the Colorado is to maintain the ecosystems of the Colorado Delta and the Gulf of California (also known as the Sea of Cortés). The construction of dams in the United States has greatly diminished the amount of freshwater flows to the gulf and resulted in highly saline conditions and increased water temperature.[51] The decline of fresh water flows into the Gulf has been shown to decrease mollusk populations,[52] and lower flows are correlated with lower abundance of profitable blue shrimp.[53]

The lack of sediment flow has also changed the geomorphic characteristics of the delta, which in turn have resulted in the loss of habitat for indigenous species, such as the now-endangered totoaba,[54] which once supported commercial fishing. If, in the future, the United States withholds all stream flow for a prolonged duration, then saline levels and water temperatures in the Gulf may change even more and consequently further degrade the local fishing industry. If the lack of food and water result in refugee migration, and if refugee migration is a national security concern, then the action to keep all of the Colorado water may exacerbate threats to the United States' security. Similarly, if the lack of food and water result in Mexico's internal instability, then, again, this action may exacerbate U.S. national security concerns. Given these risks, the decision to break the treaty may be inconsistent with the larger logic of the scenario's defensive posture. The decision to cancel the treaty becomes even more problematic when one considers how it might affect other relationships between the United States and Mexico. For example, the North American Free Trade Agreement (NAFTA) and the politically related North American Agreement on Environmental Cooperation (NAAEC) provide the means to adjudicate disputes related to environmental differences that may result from cross-border activities, including trade. Negatively affecting fishing grounds or withholding water for crop irrigation could be seen as triggering such a dispute. Given the interconnectedness of the U.S. economy—which the scenario notes—and Schwartz and Randall's definition of carrying capacity, which includes an economic dimension, would the U.S. government proceed to cancel the 1944 treaty if it could also result in the demise of NAFTA and the loss of a trading partner? The answer is perhaps, but also perhaps not. The point to be made here is that because the report does not provide an adequate framework for grappling with the complexities of state-level actions, it is difficult for the reader to assess the logic of the actions that are mentioned.

This criticism is not to say that some use of LeBlanc's work is not useful to help understand resource-related conflict; only that this one book is not, in and of itself, sufficient for the task of global political analysis about the future. Toward a more limited application of the ideas expressed in *Constant Battles,* those interested in future geopolitics might take note of LeBlanc's observation that there are parts of the globe that still function along the lines of tribes and chiefdoms, rather than as nation-states.[55] Many of these kinds of places have been identified as deserving scrutiny for national security concerns by Thomas P. M. Barnett, formerly of the Naval War College. Barnett distinguishes between states within what he calls the functioning "Core" of social and economic globalization and those of the nonintegrating "Gap," which have, for various reasons, resisted globalization.[56] Geographically, the Gap spans from northwestern South America to much of the Caribbean to most of Africa to the Middle East to Central Asia to parts of the South Pacific. The Gap is characterized by poverty, violence, repressive political systems, and low life expectancies. Also, many parts of the Gap already suffer from ecological degradation. Barnett notes that throughout the 1990s, most U.S. military deployments were to areas in the Gap. In light of LeBlanc's extensive documentation of tribes and chiefdoms functioning in the face of ecological stress, the need for intervention in these places may not come as a surprise. Following abrupt climate change, conditions in the Gap may worsen and demand even more military attention. An exploration that looks for more specific commonalities (and differences) between the work of LeBlanc and Barnett may produce a more insightful basis for scenarios that deal with this part of the world.

At the other extreme from the premodern Gap of tribal- and chiefdom-dominated regions and beyond much of the modern Core of nation-states, Robert Cooper has proposed the concept of the postmodern state that transcends national borders and that operates on the basis of openness, law, and mutual security.[57] The most notable emerging example of the postmodern state is the European Union. Japan, Cooper argues, is postmodern by inclination, given its self-imposed limits on defense activities and its multilateralist approach to foreign affairs. These places feature prominently in Schwartz and Randall's scenario, and a consideration of this newer world order could be used as a basis to grapple with possible dynamics of future actions taken by these entities. Although far from making a prediction, Cooper also suggests the possibility that the United States might come to operate like a postmodern state. If that is the case, then another avenue for scenario development would be to consider the implications of a U.S. strategy based not on diametrically opposed terms of offensive or defensive postures, but in varying degrees of openness.

LOOKING FORWARD

An inset box at the beginning of Schwartz and Randall's report titled "Imagining the Unthinkable" (Appendix One) provides a clear allusion to Herman Kahn's seminal notion of "Thinking the Unthinkable." In making this connection, the authors not only reach back to the person who can be considered the founder of security-oriented scenario studies; they also attempt to engage in his kind of inquiry into problems that are technically intricate, are socially complex, and defy simple analysis. By some measures, more than a metaphorical bridge has been made to this past work. Kahn hoped to encourage serious public discussion of national security concerns, and his many writings can be seen as providing fodder for open debate. By the same token, Schwartz and Randall's approachable scenario provides a basis for nonexperts to join an active worldwide conversation. Moreover, beyond simply enabling a common basis for discussion, Schwartz and Randall may have provided a means to vent a source of pent-up public anxiety. Perhaps the exaggerated response by the press was not (or at least not only) a crass effort to sell more newspapers but reflected that the scenario struck a collective nerve.

To be sure, there are contextual differences between Cold War thoughts on thermonuclear war and contemporary thoughts on abrupt climate change. First, discussions about the environmental dimensions of national security can be much more open than discussions of nuclear weaponry. Kahn's writings show the awareness that some details of Cold War planning had to be kept from the public, lest the Soviets learn too much about U.S. or NATO capabilities and initiatives. Today, not only are the facts of ecological conditions to be found in "open source" information, but military reconnaissance assets are now commonly used for civilian environmental monitoring activities.[58] Second, the geopolitics of the Cold War often reduced the understanding of the world to a zero-sum game in which one side's loss was the other side's gain. The national security threats posed by abrupt climate change will potentially affect every nation and every person on the planet—and for most in negative ways. Those nations that are better prepared may, in the long term, be better off, but that is not to say there will be winners and losers. Today, new frameworks to assess the interconnectedness of national and global security are being developed and should be used to understand common concerns and common solutions. Third, the fact that climate change is being discussed at all under the rubric "national security" indicates that the kinds of threats that are recognized have greatly increased since the Cold War. Reflecting this new situation, the range of specialists who can provide useful, if not much needed, insight has necessarily expanded. Today, seminars on security studies require a larger table.

These contextual differences lead to new opportunities and requirements for scenario development. Because the discussion can be open, assumptions about the future *can* be more explicitly stated. Because the discussion should seek, explore, and chart new relationships between society and the environment, the assumptions about the future *should* be more explicitly stated. And because the discussion must include people from multiple disciplines and diverse backgrounds who must talk to each other, the assumptions about the future *must* be more explicitly stated. So, what should be done next to sharpen our thinking about future climate change and national security?

Based on their efforts, Schwartz and Randall make the following recommendations to prepare the United States for abrupt climate change:

1. Improve predictive climatic models;
2. Assemble comprehensive models of climate change impact;
3. Create vulnerability metrics to understand a country's vulnerability to climate change;
4. Identify no-regrets strategies that would ensure reliable access to food and water and ensure national security;
5. Rehearse adaptive responses that would prepare for inevitable climate-driven events such as mass migration, disease, and food and water shortages;
6. Explore local implications; and
7. Explore geoengineering options that control climate.

Each of these recommendations seems a prudent step toward bettering our understanding of the forces and consequences of change. But none explicitly involves an effort to improve our grasp of the possible cultural, social, economic, and political shifts that might emerge with new environmental conditions. If we are to understand the full implications of climate change on national security, we will also need to examine potential reasons for action.

In short, the Pentagon ONA's decision to commission a scenario on abrupt climate change suggests that the military is aware of a need to expand its understanding of national security. Schwartz and Randall's study makes a contribution toward meeting that need. As discussed in this chapter, the scenario has its limitations, but it has served to raise public awareness and may yet further some level of open debate. We would hope that the Pentagon has not prematurely shelved the document but will read it with an eye toward continuing with the systematic, hard-headed work that must be done.

NOTES

1. Peter Schwartz and Doug Randall, "An Abrupt Climate Change Scenario and Its Implications for United States National Security" (Emeryville, CA: Global Business Network [GBN], 2003). Available online through the GBN Web site (www.gbn.com).

2. David Stipp, "The Pentagon's Weather Nightmare," *Fortune,* February 9, 2004, pp. 100–108.

3. Edward Ortiz, "Pentagon Report Plans for Climate Catastrophe," *Providence Journal* (Rhode Island), March 3, 2004, Section B, p. 1.

4. Mark Townsend and Paul Harris, "Now the Pentagon Tells Bush: Climate Change Will Destroy Us," *The Observer,* 22 February 2004, observer.guardian.co.uk.

5. *The Day After Tomorrow,* DVD, directed by Ronald Emmerich (Beverly Hills, CA: 20th Century Fox Home Entertainment, 2004).

6. Andrew C. Revkin, "The Sky Is Falling! Say Hollywood and, Yes, the Pentagon," *New York Times,* February 29, 2004, Section 4, p. 5.

7. Keay Davidson, "Pentagon-Sponsored Climate Report Sparks Hullabaloo in Europe, but New Ice Age Unlikely, Bay Area Authors of Study Say," *San Francisco Chronicle,* February 25, 2004, Section A, p. 2.

8. Steve Curwood, "Abrupt Climate Change," *Living on Earth,* March 5, 2004, www.loe.org.

9. Revkin, p. 5.

10. National Research Council (U.S.), Committee on Abrupt Climate Change, *Abrupt Climate Change: Inevitable Surprises* (Washington, DC: National Academy Press, 2002).

11. Peter Schwartz, *Inevitable Surprises: Thinking Ahead in a Time of Turbulence* (New York: Gotham Books, 2003).

12. Peter Schwartz, Peter Leyden, and Joel Hyatt, *The Long Boom: A Vision for the Coming Age of Prosperity* (Reading, MA: Perseus Books, 1999).

13. Peter Schwartz, *The Art of the Long View: Paths to Strategic Insight for Yourself and Your Company* (New York: Currency Doubleday, 1996).

14. Clayton Roberts, *The Logic of Historical Explanation* (University Park: Pennsylvania University Press, 1996).

15. Kenneth Burke, *A Grammar of Motive* (Berkeley: University of California Press, 1945).

16. Pierre Wack, "Scenarios: Uncharted Waters Ahead," *Harvard Business Review* 65 (September-October 1985): pp. 72–79; this note p. 77. Schwartz (1996), pp. 101, 108–14.

17. A. J. P. Taylor, *War by Time-Table: How the First World War Began* (London: Macdonald & Company, 1969).

18. Thomas R. Karl and Kevin E. Trenberth, "Modern Global Climate Change," *Science* 302 (December 5, 2003): pp. 1719–23.

19. Intergovernmental Panel on Climate Change, *Climate Change 2001: Synthesis Report—Summary for Policymakers* (Wembley, UK: IPCC Plenary XVIII, September 24–29, 2001), p. 5.

20. Mort Webster, "Communicating Climate Change Uncertainty to the Policy-Makers and the Public," *Climatic Change* 61 (November 2003); pp. 1–8.

21. Hillel J. Einhorn, "Expert Judgment: Some Necessary Conditions and an Example," in *Judgment and Decision Making: An Interdisciplinary Reader,* ed. Terry Connolly, Hal R. Arkes, and Kenneth R. Hammond, 2nd ed. (Cambridge, UK: Cambridge University Press, 2000), pp. 324–35; also, Hillel J. Einhorn, "Expert Judgment: Some Necessary Conditions and an Example," *Journal of Applied Psychology* 59:5 (1974): pp. 562–71.

22. D. A. Clark, "Sources or Sinks? The Responses of Tropical Forests to Current and Future Climate and Atmospheric Composition," *Philosophical Transactions of the Royal Society of London Series B—Biological Sciences* 359 (March 29, 2004): pp. 477–91.

23. John Reilly, Peter H. Stone, Chris E. Forest, Mort D. Webster, Henry D. Jacoby, and Ronald G. Prinn, "Uncertainty and Climate Change Assessments," *Science* 293 (July 20, 2001): pp. 430–33.

24. M. Granger Morgan and David W. Keith, "Subjective Judgments by Climate Experts," *Environmental Science and Technology* 29 (October 1995): pp. 468A–76A.

25. Anthony G. Patt and Daniel P. Schrag, "Using Specific Language to Describe Risk and Probability," *Climatic Change* 61 (November 2003): pp. 17–30.

26. Stephen H. Schneider, "What Is 'Dangerous' Climate Change?" *Nature* 411 (May 3, 2001): pp. 17–19.

27. Richard B. Stewart and Jonathan B. Wiener, "Practical Climate Change Policy," *Issues in Science and Technology* 20 (Winter 2004): pp. 71–78.

28. George W. Bush, "Remarks by President Bush on Global Climate Change" (Washington, DC: The White House, June 11, 2001), www.state.gov/g/oes/rls/rm/4149.htm.

29. National Research Council (U.S.), Committee on Abrupt Climate Change, p. 41.

30. Richard H. Ullman, "Redefining Security," *International Security* 8 (Summer 1983): pp. 129–53; Jessica Tuchman Mathews, "Redefining Security," *Foreign Affairs* 68 (Spring 1989): pp. 162–77; Norman Matthews, "Environment and Security," *Foreign Policy* 74 (Spring 1989): pp. 23–41.

31. Geoffrey D. Dabelko and David D. Dabelko, "Environmental Security: Issues of Conflict and Redefinition," *Environmental Change and Security Project Report* 1 (Spring

1995): pp. 3–13; Peter H. Gleick, "Environment and Security: The Clear Connections," *The Bulletin of Atomic Scientists* 47 (April 1991): pp. 16–21.

32. Dabelko and Dabelko.

33. Franklyn Griffiths, "Environment in the US Security Debate: The Case of the Missing Arctic Waters," *Environmental Change and Security Project Report* 3 (Spring 1997): pp. 15–28.

34. Jon Barnett, "Security and Climate Change," *Global Environmental Change* 13 (April 2003): pp. 7–17.

35. U.S. President (1989–1993, George H. W. Bush), *National Security Strategy* (Washington, DC: U.S. Government Printing Office, 1991), p. 22.

36. U.S. President (1993–2001, William J. Clinton), *A National Security Strategy of Engagement and Enlargement* (Washington, DC: U.S. Government Printing Office, 1994), p. 1.

37. Sherri Wasserman Goodman, "The Environment and National Security," Remarks to National Defense University (Washington, DC, August 8, 1996), www.loyola.edu/dept/politics/intel/Goodman.html.

38. Daniel Deudney, "Environment and Security: Muddled Thinking," *The Bulletin of Atomic Scientists* 47 (April 1991): pp. 22–28.

39. Barry Buzan, Ole Weaver, and Jaap de Wilde, "Environmental Security and Societal Security," *Working Papers* No. 10 (Copenhagen, Denmark: Center for Peace and Conflict Research, 1995).

40. Schwartz and Randall, p. 15.

41. Donella H. Meadows, *Limits to Growth: A Report for the Club of Rome's Project on the Predicament of Mankind* (New York: Universe Books, 1972).

42. John C. Woodwell, "A Simulation Model to Illustrate Feedbacks among Resource Consumption, Production, and Factors of Production in Ecological-Economic Systems," *Ecological Modelling* 112 (October 15, 1998): pp. 227–47; G. O. Barney, "The Global 2000 Report to the President and the Threshold 21 Model: Influences of Dana Meadows and System Dynamics," *System Dynamics Review* 18 (Summer 2002): pp. 123–36. The original Limits to Growth idea was also reconsidered with an expanded model in Donella H. Meadows, Dennis L. Meadows, and Jordan Randers, *Beyond the Limits: Confronting Global Collapse, Envisioning a Sustainable Future* (Post Mills, VT: Chelsea Green Publishing Company, 1992).

43. H. S. D. Cole, Christopher Freeman, Marie Jahoda, and K. L. R. Pavit eds., *Models of Doom: A Critique of the Limits to Growth* (New York: Universe Books, 1973).

44. George E. P. Box, "Robustness in the Strategy of Scientific Model Building," in *Robustness in Statistics: Proceedings of a Workshop,* ed. Robert L. Launer and Graham N. Wilkinson (New York: Academic Press, 1979), pp. 201–36; this note p. 202.

45. Personal telephone conversation with Steven A. LeBlanc, March 5, 2004.

46. Steven A. LeBlanc with Katherine E. Register, *Constant Battles: The Myth of the Peaceful, Noble Savage* (New York: St. Martin's Press, 2003).

47. LeBlanc with Register, pp. 38–39.

48. Schwartz and Randall, p. 16.

49. LeBlanc with Register, p. 166.

50. Ibid., p. 197.

51. Robert G. Varady, Katherine B. Harkins, Andrea Kaus, Emily Young, and Robert Merideth, " … to the Sea of Cortés: Nature, Water, Culture, and Livelihood in the Lower Colorado River Basin and Delta—An Overview of Issues, Policies, and Approaches to Environmental Restoration," *Journal of Arid Environments* 49 (September 2001): pp. 195–209; this note, p. 205.

52. Carlie A. Rodriguez, Karl W. Flessa, and David L. Dettman, "Effects of Upstream Diversion of Colorado River Water on the Estuarine Bivalve Mollusc *Mulinia coloradoensis,*" *Conservation Biology* 15 (February 2001); pp. 249–58.

53. Eugenio Alberto Aragón-Noriego and Luis Eduardo Calderón-Aguilera, "Does Damming the Colorado River Affect the Nursery Area of Blue Shrimp *Litopenaeus stylirostis* (Decapoda: Penaeidae) in the Upper Gulf of California?" *Revista de Biologia Tropical* 48 (December 2000): pp. 867–71.

54. J. D. Carriquiry and A. Sánchez, "Sedimentation in the Colorado River Delta and Upper Gulf of California after Nearly a Century of Discharge Loss," *Marine Geology* 158 (June 1999): 125–45.

55. LeBlanc with Register, p. 116.

56. Thomas P. M. Barnett, *The Pentagon's New Map: War and Peace in the Twenty-First Century* (New York: G. P. Putnam's Sons, 2004); For an abridged discussion of this argument, see Thomas P. M. Barnett, "The Pentagon's New Map," *Esquire,* March 2003, pp. 174–79, 227–28.

57. Robert Cooper, *The Breaking of Nations: Order and Chaos in the Twenty-First Century* (New York: Atlantic Monthly Press, 2003).

58. Griffiths, p. 18.

— 3 —

Zombie Concepts and Boomerang Effects: The Shifting Structure of Strategy and Security

Climate change is the most severe problem that we are facing today—more serious even than the threat of terrorism.

Sir David King
Chief Scientific Advisor to H. M. Government
and Head of the Office of Science and Technology, Great Britain
"Climate Change Science: Adapt, Mitigate, or Ignore?" Science,
January 9, 2004

The issue of climate change and its impact on humankind—and the security that affects humankind—requires, we argue, a more focused, more nuanced, more strategic approach. Fundamentally, we must begin with a sense, if not total understanding, of how the entire security landscape before us is shifting. Building on the background to the abrupt climate change scenario we introduced in chapter 2, this chapter begins to address how the security structure has radically shifted since the fall of the Berlin Wall in 1989. Following from there, in chapter 4 we consider how the mind-sets—or "mental maps"—of decision makers drive the willingness (or, more appropriately, the unwillingness) to take on policies that would respond to potential outcomes. In chapters 5 and 6 we address how the "new" security must be considered in light of the extraordinary changes that are occurring. Finally, we conclude with a set of observations and still troubling questions that remain, as well as offer some pathways that might lead to viable responses that could improve resilience, our ability to adapt, or, at least, help mitigate more negative outcomes.

We begin then with two metaphors: the "zombie concept" and the "boomerang effect." The zombie concept—sometimes called the "zombie category"—stems from the work of German sociologist Ulrich Beck. According to Beck, zombie concepts belong to a time when the approach to understanding not only security, but also one's place in the global order, rested on an identity that was "nation-state centered."[1] Beck—as well as many others—argues that we have now entered an age of true interlinked, interdependent globalization, involving a multidimensional process of change that has irrevocably changed the social and world order, and the places and function of states within that order. In other words, we are living in an industrial (or preindustrial) society *organized* according to nation-states, but we are already no longer living in it.[2] We are beyond the nation-state. We live in the "world risk society," where common threats and common opportunities abound.

Yet though the term *globalization* remains vague—something better sensed than perfectly comprehended—there is often attached to the concept and the process the recognition of *flows,* which involve not only economic but also cultural, sociological, religious, historical, political, ecological, and economic exchanges. The popular *New York Times* columnist Thomas Friedman has suggested, for example, that globalization "is the integration of virtually everything with everything." Such less-than-helpful definitions help explain why *globalization* is a frequently used and frequently abused term—and why President George H. W. Bush often referred to the phenomenon as "globaloney."

As we use the term here, globalization involves these disparate and complex flows in the vast networking of financial transactions, rising expectations in standards of living, and increasingly common forms of (largely democratic) state and regional identity and governance. (Equally, and as we continue to emphasize throughout this work, the aspect of climate change and its shared "effects" suggests that the burdens and the impacts will be shared as well.)

Before moving to a general consideration of security therefore, we would like to introduce a second controlling metaphor—that of the "boomerang effect." To best explain the intended significance of this term, we offer a simple, though personal, anecdote. One of the authors (Liotta) grew up in Australia. One weekend in Queensland, exploring several opal mines, his father elected to try out a new killer boomerang. The weapon was heavy enough to be lethal yet light enough to supposedly return when thrown. And sure enough, with the family gathered in an open field, his father heaved a mighty toss and the weapon worked as advertised. To the family's general awe and amazement, the boomerang arced several hundred meters into the air; to their horror, the boomerang then started to return. Without hesitating, his father, two brothers, and he fled to all four quadrants of the compass. His mother, frozen in place, caught the full impact of the massive boomerang. She has never truly forgiven any of them.

For the authors, the significance of the boomerang effect lies in the emerging understanding that aspects of nontraditional security issues that have long plagued the so-called developing world could increasingly affect the policy decisions and future choices of powerful states and world leaders as well. As disparate as these nontraditional issues may be—whether linked to climate change, resource scarcity, declining productivity, or transnational issues of criminality and terrorism—the developed world is now confronted with similar, human-centered vulnerabilities that had often been present previously only in the context of nontraditional challenges for developing regions.

In short, we may need to worry less about focusing on protecting the "state" and more on protecting "individual" citizens, which means protection of individual rights and liberties as well as the way of life that most have become accustomed to in the developed world. The irony in this claim, of course, is that many proponents for development in some of the poorest states have long argued that the focus on the individual—rather than on sustaining the power base of the state—is the best guarantee for long-term stability, prosperity, and security.

The implications of the changing security landscape for the analyst and policy maker are therefore potentially profound. In essence, we may be witnessing a boomerang effect in which we must focus *both* on aspects of national security, in which military forces may continue to play a preeminent role, and on human security, in which nontraditional security issues predominate. Thus, we may well witness renewed focus on failed or failing states, epidemiology (such as AIDS or H5N1), environmental stress, resource scarcity and depletion, drugs, terrorism, small arms, inhumane weapons, cyber-war, and narco-trafficking. Predominant among these issues, of course, is climate change in an apparent time when "technologically advanced society could choose, in essence, to destroy itself, but that is now what we are in the process of doing."[3]

At this point, nonetheless, we must stress that climate change, of itself, is not an effect "in isolation"; rather, as the alarmist Schwartz and Randall scenario suggests, there is a complex interdependence and a dynamic interplay among a host of dependent and independent variables that cannot fully be predicted or understood. In its way, this complex interaction approaches what was once known as chaos theory. And yet, as disparate as these nontraditional security aspects indeed are, they all will (in one form or another and in multiple geopolitical contexts) increasingly have influence on future strategic relationships and decisions. The issue truly is not one of "hard" traditional security (often based on state-to-state power relationships) or "soft" nontraditional security (that can involve multiple transnational aspects). The future will require decision makers in both the developing and developed world to focus on broad—and increasingly broadening—understandings of the meaning of security. Focusing on one aspect of security at the expense or detriment of another aspect, nevertheless, may well cause us to be "boomeranged" by a poor balancing of ends and means in a radically changed security environment. Before proceeding further, therefore, we may need to ask and to answer what it is we mean when we say "security."

WHAT IS SECURITY?

In the classical sense, security—from the Latin *securitas*—refers to tranquility and freedom from care, or what Cicero termed the absence of anxiety upon which the fulfilled life depends. Notably, numerous governmental and international reports that focus on the terms "freedom from fear" and "freedom from want" emphasize a pluralist notion that security is a basic, and elemental, need.

Conversely, the stable state extended upward in its relations to influence the security of the overall international system. The overall system grew to promote security by supporting the stability of states. Individual security, stemming from the liberal thought of the Enlightenment, was also considered both a unique and collective good. Adam Smith, for example, in *The Theory of Moral Sentiments,* mentions only the security of the sovereign, who possesses a standing army to protect him against popular discontent, and is thus "secure" and able to allow his subject the liberty of political "remonstrance." By contrast, M. J. de Condorcet's argument, in the late eighteenth century, suggested that the economic security of individuals was an essential condition for political society; fear—and the fear of fear—were for Condorcet the enemies of liberal politics.

Individual security, stemming from the liberal thought of the Enlightenment, was also considered both a unique and collective good. Moreover, despite the abundance of theoretical and conceptual approaches in recent history, the right of states to protect themselves under the rubric of "national security" and through traditional instruments of power (political, economic, and especially military) has never been directly, or sufficiently, challenged. The *responsibility,* however, for the guarantee of the individual good—under any security rubric—has never been obvious.

From this rather general—and quite European—understanding of security, the human security concept centers on a concentration on the individual (rather than the state) and that individual's right to personal safety, basic freedoms, and access to sustainable prosperity.[4] In ethical terms, human security is both a system and a systemic practice that promotes and sustains stability, security, and progressive integration of individuals within their relationships to their states, societies, and regions. In abstract but understandable terms, human security allows individuals the pursuit of life, liberty, and the pursuit of both happiness and justice.

Clearly, one could find little to argue with in these principles. There are problems, nonetheless. On the one hand, all security systems are not equal—or the same. Moreover, all such systems collectively involve codes of values, morality, religion, history, tradition, and even language. Any system that enforces, as it were, human security inevitably collides with conflicting values—which are not synchronous or accepted by all individuals, states, societies, or regions.

On the other hand, in the once widely accepted realist understanding, the state was the sole guarantor of security. (In the contemporary geopolitical landscape, realists continue to assert that security necessarily extends downward from nations to individuals; conversely, the stable state extended upward in its relationship to other states to influence the security of the international system. This broadly characterizes what is known as the anarchic order.) For Thomas Hobbes, the classic state-centered realist, an individual's insecurity sprang from a life that was "solitary, poor, nasty, brutish and short."[5] The state protected the individual from threats, whether these threats came at the hands of a local thief or from an invading army. For this protection, the citizen essentially relinquished individual rights to the state, as the state was the sole protector.

Thus, in contrast to principles embedded in documents such as the American Declaration of Independence and the U.S. Constitution, security always trumped liberty. Clearly in an age when terrorism and extremist violence are constant challenges, and when legislation such as the Patriot Act and individual surveillance continue unabated in what are considered "open" societies, the conflict between collective/individual security and individual liberty remains. Indeed, Benjamin Franklin's adage remains an uncomfortable dilemma even today: Those who give up their personal liberty for increased security deserve neither.

It does seem significant that aspects of "nontraditional" security issues that have long plagued the so-called developing world—issues that include environmental degradation, resource scarcity, epidemiology, transnational issues of criminality, and terrorism—can increasingly affect the policy decisions and future choices for powerful states and world leaders as well. The 1994 United Nations Development Programme (UNDP) report, for example, was an attempt to recognize a conceptual shift that needed to take place:

The concept of security has for too long been interpreted narrowly: as security of territory from external aggression, or as protection of national interests in foreign policy or as global security from the threat of nuclear holocaust. It has been related to nation-states more than people. . . . Forgotten were the legitimate concerns of ordinary people who sought security in their daily lives. For many of them, security symbolized protection from the threat of disease, hunger, unemployment, crime [or terrorism], social conflict, political repression and environmental hazards. With the dark shadows of the Cold War receding, one can see that many conflicts are within nations rather than between nations.[6]

In 2003, the UN Commission on Human Security expanded this concept to include protection for peoples suffering through violent conflict, for those who are on the move whether out of migration or in refugee status, for those in postconflict situations, and for protecting and improving conditions of poverty, health, and knowledge.[7]

One would think that with the fall of the Berlin Wall it should have become clear that despite the macro-level stability created by the East-West military balance of the Cold War, citizens were not safe. They may not have suffered from outright nuclear attack, but they were being killed by the remnants of proxy wars, the environment, poverty, disease, hunger, violence, and human rights abuses. Ironically, the faith placed in the realist worldview, and the security it provided, masked the actual issues threatening the individual. The protection of the person was all too often negated by an overattention on the state. Allowing key issues to fall through the cracks, "traditional security" failed at its primary objective: protecting the individual.

This new type of instability led to a challenge to the notion of traditional security by such concepts as cooperative, comprehensive, societal, collective, international, and human security.

The 1994 UNDP report therefore attempted to argue that freedom from chronic threats such as hunger, disease, and repression (which require long-term planning and development investment) as well as the protection from sudden disasters (which require often immediate interventions of support from outside agents) required action. Thus the UNDP offered seven "nontraditional" security components:[8]

1. Economic security: The threat is poverty; vulnerability to global economic change
2. Food security: The threat is hunger and famine; vulnerability to extreme climate events and agricultural changes
3. Health security: The threat is injury and disease; vulnerability to disease and infection
4. Environmental security: The threat is resource depletion; vulnerability to pollution and environmental degradation
5. Personal security: The threat is from violence; vulnerability to conflicts, natural hazards, and "creeping" disasters
6. Community security: The threat is to the integrity of cultures; vulnerability to cultural globalization
7. Political security: The threat is of political repression; vulnerability to conflicts and warfare.

In this conceptual approach to "new" security, the overarching focus is on the individual. And, even as these components fractured security into separate identities, the core remained a focus on the human citizen. In pragmatic ways, the broad conceptualization of security is revolutionary—and quite different from a traditional, state-centric view of security.

A year later, Emma Rothschild usefully depicted how security has changed horizontally, vertically, and on multiple axes. Beginning with the state, she described security as brought down to the individual; up to the international system or supranational physical environment; and across (broadening) from a focus on the military to include the environment, society, and economy. Moreover, she argued that the responsibility to ensure security diffused in all directions to include local governments, international agreements, nongovernmental organizations (NGOs), public opinion, forces of nature, and the financial market. Although not an explicit definition, this conceptualization provides an example of both how narrow the traditional paradigm has been, as well as how complex the expansion of the concept can become.[9]

Essentially, states and regions, in a globalized context, can no longer afford to solely emphasize national security issues without recognizing that abstract concepts such as values, norms, and expectations also influence both choice and outcome. Societies, whether in the emerging world or in the developed world (admittedly, a rather arrogant term), are increasingly witnessing an unfolding tension: Whereas there are expectations that states have for protecting their citizens, citizens increasingly hold their states accountable.

Within this debate on security, there emerged an increased focus on the rights of the individual. This debate has led to intriguing possibilities and, most definitely, uncertain outcomes. It remains unclear, however, whether an ethical and collective policy to support human security will be the focus of most states in the future—or whether any such policy could logically be delinked from a systematic association with power and powerful states.

In such an environment, to avoid the boomerang effect, decision makers and policy analysts will, again, need to focus on multiple aspects, axes, and levels of security. Despite the "clean" distinctions made in Table 3.1, there are dangers in too closely following the precepts of one security concept at the expense of another.[10]

Some brief explanation of the concepts used in this graphic might prove useful. In essence, the distinctions move from a "top-down" global emphasis to a "bottom-up" individual focus.

Environmental security emphasizes the sustained viability of the ecosystem, while recognizing that the ecosystem itself is perhaps the ultimate weapon of mass destruction. In 1556 in Shensi province, for example, tectonic plates shifted and by the time they settled back into place, 800,000 Chinese were dead. Roughly 73,500 years ago, a volcanic eruption in what is today Sumatra was so violent that ash circled the earth for several years, photosynthesis essentially stopped, and the precursors to what is today the human race amounted to only several thousand survivors worldwide.[11] Yet from an alternative point of view, mankind itself is the ultimate threat to the ecosystem. Thus, from a radically extreme perspective, elimination of humanity proves the ultimate guarantee of the ecosystem's survival.

National security represents the traditional understanding of security, to include the protection of territory and citizens from external threats—from other states, and, more recently, "stateless" actors (which range from NGOs to terrorist

Table 3.1
Alternative "Security" Concepts

Traditions and Analytic Bases	Security Type	Specific Security Focus	Specific Security Concern	Specific Security Hazards (Threats/ Vulnerabilities)
Non-traditional	Ecological Security	The Ecosystem	Global sustainability	Mankind: through resource depletion, scarcity, war, and ecological destruction
Traditional, realist-based	National Security	The State	Sovereignty, Territorial integrity	Challenges from other states (and "stateless" actors)
Traditional & non-traditional realist- and liberal-based	"Embedded" Security	Nations, Societal groups, Class/economic, Political action committees/interest groups	Identity/Inclusion, Morality/Values/ Conduct, Quality of life, Wealth distribution, Political cohesion	States, Nations, Migrants, Alien culture
Non-traditional, liberal/ Marxist-based	Human Security	Individuals, Mankind, Human rights, Rule of Law, Development	Survival, Human progress, Identity and governance	The state itself, Globalization, Natural catastrophe and change

networks). Hyper-emphasis on state security, especially in the emergence of "homeland security," affects the two following concepts of security, especially regarding the practice of individual liberties and the freedom to participate openly in civil society.

Embedded security is not synonymous with the more commonly used term "societal security." Rather, embedded security is somewhat symbiotically (perhaps parasitically) linked to other security concepts. It often represents the narrow interests of specific communities, nations, or political action groups within a state. In its extreme form, it can lead to social stratification, the fracturing of "common" interests, and xenophobia. Samuel P. Huntington, all normative judgments aside, focused on such embedded security groups in his 2004 work, *Who Are We? Challenges to America's National Identity.* Moreover, exit polling data from the American presidential election in November 2004 reports that the single most important criterion for selection of a president (23 percent of all those polled) was "moral values"—more important than how a president would deal with economics, foreign policy, terrorism, the war in Iraq, education, or health care. Although the term "moral values" is loaded with ambiguity, it represents a form of parasitic embedded security. In the example of the American election, a specific interest group—Christian Evangelicals—most associated with "moral values," also most influenced the election's outcome.

Human security remains a controversial term. Other than a common agreement on the focus on the individual, the concept of human security is still emerging.[12] In simple terms, the United Nations Commission on Human Security defines *human security* as the protection of "the vital core of all human lives in ways that enhance human freedoms and fulfillment."[13] In the September 1999 issue of *Security Dialogue,* Astri Suhrke pointed to a fundamental bifurcation that human security as conceptual approach and policy principle continues to suffer from: Is it related more to long-term "human development," such as was suggested in the 1994 United Nations *Human Development Report,* or (as a security issue) does it constitute a principle of intervention during immediate crisis, such as Rwanda in 1994, Kosovo in 1999, or even Iraq in 2003? The answer to either question is "Nes"—a little bit of no and a little yes. For the sake of simple argument, we define human security as "protecting people."

Thus, while some have argued that there may be a growing convergence between what was traditionally called "national security" and the still-developing concept of "human security," there appears to be an even more powerful counterargument in which the opposite trend is apparent.[14] In interventions as disparate as Somalia in 1993 to Liberia (at various stages of disintegration) to the Balkans and Iraq in 2003, there has emerged an overt increase in American and British hegemonic behavior, accompanied by an uneven commitment to issues involving human security. While Prime Minister Tony Blair could speak of "universal values" and President G. W. Bush proclaimed that "Freedom is the non-negotiable demand of human dignity," foreign policy choices regarding intervention were almost exclusively made when such choices satisfied the narrow, selfish, and direct "national security interests" of more powerful states. If such choices also satisfied certain aspects of human security, then all the better.

Yet, as the blatant international failure in 1994 in Rwanda to do anything illustrates—other than a collective international decision to do *nothing*—human security is hardly proving to be the trump card of choice in decisions by states to intervene in the affairs of other states, to include violating traditionally respected rights of sovereignty. In other words, taken to extreme forms, both human security and national security can be conceptually approached as antagonistic rather than convergent identities. Each, in its exclusive recognition, remains problematic.

So-called ethical practice in foreign policy that acts on the behalf of individual citizens, for *any* state—not just the United States—is most accommodated (and accommodating) when such actions "mesh" with achieving the ends of more traditional national interests of the more powerful state. As a basis for international action, we have yet to achieve a consensus on what constitutes "international interests," who should support it, and who should uphold it.

Human security, both as conceptual approach and policy principle, thus rests uncomfortably on the horns of a dilemma. Although the effort to promote human security in the arena of "high politics" on the part of the Canadian and Norwegian governments since the 1990s is well-known, there is a tempting sense of proselytizing righteousness as well. Such so-called middle power states, after all, can exercise significant moral clout by emphasizing that the rights of the individual are at least as important as protecting the territorial and sovereign integrity of the state. Yet when larger powers, particularly those with significant militaries (such as the United States or the United Kingdom) advocate similar positions, it is their overwhelming power that is recognized, respected, and resented.

On the one hand, what is perceived as the "moral clout" of the middle power is sensed as "hegemony unbridled" when it is emphasized in an attempted similar fashion by major powers. On the other hand, when actions taken in the name, or in the principled following, of human security do occur, they are often inextricably linked to issues that are embedded in the more traditional concepts of "national security" and protection of the state. Idealism thus becomes enmeshed in realism; actions taken on behalf of the powerless are determined only by the powerful.[15] As George Packer notes, referring to both the Kosovo and Iraq interventions, and the state-declared reasons for them: "The [Bush] Administration has given idealism a bad name, and it will now take years to rescue Václav Havel from Paul Wolfowitz."[16]

Undoubtedly, increasing numbers now speak out on behalf of what the International Commission on Intervention and State Sovereignty has termed the "responsibility to protect": the responsibility of some agency or state (whether it be a superpower such as the United States or an institution such as the United Nations) to enforce the principle of security that sovereign states owe to their citizens. But there is a dark side of this proposition, of course: The "responsibility to protect" also means the "right to intervene." In the topology of power, dominant states will intervene at the time and place of their choosing.

Overall, we argue, a conceptual shift in the approaches of states and international organizations has begun to take place. The cultural and political philosopher J. Peter Burgess has aptly summarized, for example, a major European shift regarding the concept of security since the end of the Cold War:

In *New & Old Wars* Mary Kaldor argues that a new type of organized violence has developed, beginning in the 1980s and 1990s, as one aspect of the globalized era. The new wars are, according to Kaldor, characterized by a blurring of the distinctions between war, organized crime, and wide-scale violations of human rights. In contrast to the geopolitical goals of earlier wars, the new wars are about identity politics. Kaldor argues that in the context of globalization, ideological and territorial cleavages of an earlier era have increasingly been supplanted by an emerging political cleavage between cosmopolitanism, based on inclusive, universalist multicultural values, and the politics of particularistic identities. The evolution of the European Defense and Security Policy has evolved in the shadow of this mutation. A European culture with dubious historical reputation for cosmopolitanism is being thrust upon the global stage at the very moment when its geopolitical concepts are poised on the precipice of desuetude. With Solana's Thessaloniki Summit document "A Secure Europe in a Better World" the European community of values is being transformed into a security community.[17]

Our argument, therefore, in examining the unfolding security realities that have emerged in recent history, is to focus on multilevel, multireferent, and interdependent aspects of security. If we fail to do this, the boomerang will come arcing back toward us. Indeed, we may well never know what, eventually, will hit us.

WHAT WE SHOULD TALK ABOUT WHEN WE TALK ABOUT SECURITY

Although security—as basic concept—is frequently considered in the study and analysis of international relations and strategy, military history, and national policy decisions, its essential meaning might better be widely debated than agreed on. Commonly considered a basic concept in policy and academic debates, security at the national and subnational level is in reality an ephemeral quantity, its definition is in large measure a reflection of the perspectives and physical situations of the student and analyst. Thus academics and analysts raised in a particular school—the realist school that emphasizes power relationships between and among states being the most influential at least in government circles—tend to interpret events with the blinders that the school's focus provided. That realism has stood the test of time accounts for the skepticism that statesmen and many scholars have toward nontraditional, wider definitions that would include many of the issues we raise in this book. Although we are partially sympathetic to realist concerns that broad definitions lead to prescriptions for the misuse of military power, or to the underestimation of the role that military forces should and must play in world affairs, we also recognize that many of the security issues we consider here do not—and cannot—fit within a state-centric, power-driven level of analysis.

Allowing for such intellectual carelessness leads all those who study and practice security into dangerous territory, therefore. By couching emerging nontraditional concepts such as environmental security and human security solely on their relationship to potential or real threats, most often within a topology of power—and by using language that is inadequate to the often nuanced and almost always complex dynamics of such emerging identities— immediately captures such concepts as hostage to traditional state-centered, national security paradigms.

To be blunt, there have been specific reasons for those intending to affect the debate for using such (perhaps even unintentional) strategies: Doing so both makes the topic accessible for decision makers and provides a basis for determining present and future policy. Most often such decision makers only conceive of security concepts in power-dominant, state-centric mind-set. There is a hazard, nevertheless, of adding the term "security" to either environmental or human-centered concerns. Conflating national security, human security, and environmental security all within a distinct conceptual framework, furthermore, is not only precarious; it also entails potential hypocrisy. Although admittedly a contentious claim, sure to provoke debate, the time has come to recognize some hard certainties in an increasingly complex, uncertain world.

RECOGNIZING THREATS, DISTINGUISHING VULNERABILITIES

Not all security issues involve threats; rather, the notion of vulnerabilities is as serious to some peoples—and some regions—as the familiar threat metaphor of armies massing at the borders, or barbarians at the gates. Equally, not all security issues need be directly linked to violence. One of the most serious security issues of the future, for example (particularly when considered in its relationship to climate change and human impact) is rapid urbanization of the Middle East and Southwest, South, and Southeast Asia.

Thus, an important acknowledgment should emerge here: Those who form policy and make critical decisions on behalf of states and of peoples must, ever increasingly, focus on aspects of traditional national security, in which military forces will likely continue to play a preeminent role, as well as human security, in which nontraditional security issues predominate—in which other approaches will take center stage. If such a premise proves true, and in a future where both "hard" and "soft" security will matter, those involved in policy decisions (and those affected by such decisions) will increasingly need to focus on aspects of *both* threats and vulnerabilities. There is a crucial need, then, to recognize the difference between these two categories.

A threat is both an external and internal cause of harm, that is: *identifiable, often immediate, and requires understandable response.* Military force, for example, has traditionally been sized against threats: to defend a state against external aggression, to protect vital national interests, and to enhance state security. (The size of the U.S. and USSR nuclear arsenals during the Cold War made perhaps more sense than today because the perceived threat of global holocaust in the context of a bipolar, ideological struggle was far greater then. Force structures were linearly related in the sense that their sizes reflected generally accepted assessments in each country of the other's hostility and military capability. In fact, the U.S. budget processes more or less require the identification of credible threat scenarios and attempts to keep the level of military investments in rough equivalence to the magnitude of the threat scenarios.)

Equally, the ability to apply force—not just the application of it—matters in recognizing threat challenges. As such, force options, traditionally, have included a range of responses—from deterrence to intervention to preemptive strike. All have been sized, shaped, designed, and budgeted for response to threats. A threat, in short, is either *clearly visible or commonly acknowledged.*

Vulnerabilities are less clear and less immediate. Often vulnerabilities are signaled, at best, by suggestive indicators; in some instances—a critical point here—a vulnerability, bound by its often multiple linkages to other complex factors, might not even be recognized. In the broadest understanding, vulnerabilities may not even be understood—which can be maddeningly frustrating for decision makers. When it is recognized, a vulnerability often remains only *an indicator, often not clearly identifiable, often linked to a complex interdependence among related issues, and not always suggesting a correct or even adequate response.*

Although climate change, disease, hunger, unemployment, crime, social conflict, criminality, narco-trafficking, political repression, and environmental hazards are at least somewhat related issues and do affect security of states and individuals, the best response to these related issues, in terms of security, is not at all clear. Canada, for example, has emphasized the relevance of human and environmental security to "high politics," and attempted to restructure its armed forces to meet these challenges. Yet the relevance of military state-centered forces to address or "solve" non–state-centered issues is questionable.

A vulnerability can also be both internal and external in exerting complex influence. Bohle, O'Brien, and Vogel have addressed this "double bind" of vulnerability in analyses addressing environmental change and its impact on human security. Bohle, for example, presents a simple framework for assessing these internal/external phenomena across varying levels of analysis.[18]

Figure 3.1
The Double Structure of Vulnerability

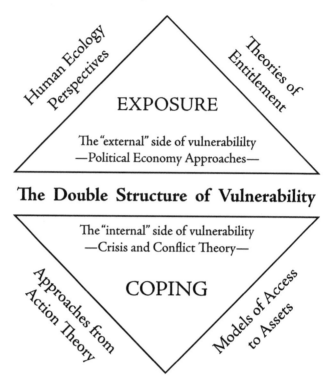

Vogel and O'Brien suggest that examining and assessing vulnerability is both relevant and applicable to policy issues concerning human security; vulnerability approaches can also identify regions and peoples at risk within the seven categories identified in the 1994 UNDP report: economic security, food security, health security, environmental security, personal security, community security, and political security.

Moreover, a vulnerability (unlike a threat) is *not clearly perceived, often not well understood, and almost always a source of contention among conflicting views.* Compounding the problem, the *time* element in the perception of vulnerability must be recognized. Some suggest that the core identity in a security response to issues involving human or environmental security is that of recognizing a condition of *extreme vulnerability.* Extreme vulnerability can arise from living under conditions of severe economic deprivation, when there are victims of natural disasters, and when there are those who are caught in the midst of war and internal conflicts. Long-term human *development* attempts thus make little to no sense and offer no direct help. The situation here, to be blunt, is not one of sustainability but of rescue.

R. H. Tawney, describing rural China in 1931, described the extreme vulnerability among peasants through a powerful image: "There are districts in which the position of the rural population is that of a man standing permanently up to the neck in the water, so that even a ripple is sufficient to drown him."[19] In such instances, the need for intervention is immediate.

Admittedly, suppositions here that insist on a distinction between threat and vulnerability become somewhat suspect in the so-called Age of Terror. Although no one doubts that certain states and actors are under threat from al Qaeda and Jemaah Islamiyah, the shadowy nature of such loosely grouped networks defies the traditional sense of threat. Loose terrorist networks often display the following characteristics: the facility to operate effectively as a lateral (and noncentralized) network, the ability to learn, the capacity to anticipate, and the capability to "self-organize" or reconstitute after they have been struck.[20] As such, these networks operate on the fault line between threat and vulnerability, and too narrow a focus on either threat or vulnerability will only lead to frustration—and failure.

UNDERSTANDING "CREEPING VULNERABILITY"

To be sure, there are cases of long-term vulnerability in which the best response is uncertain. We term these "problematic conditionalities"—which are most difficult for policy analysts and decision makers, often driven by crisis response rather than the needs of long-term strategic planning—*creeping vulnerabilities*. Given the uncertainty, the complexity, and the sheer non-linear unpredictability of creeping vulnerabilities, the frequent—and classic—mistake of the decision maker is to respond with the "gut reaction": the intuitive response to situations of clear ambiguity is, classically, to *do nothing at all*. The more appropriate response is to take an adaptive posture; to avoid the impulse to act purely on gut instinct; and to recognize what variables, indicators, and analogies from past examples might best inform the basis of action.[21]

Yet, to be clear, avoiding disastrous long-term impacts of creeping vulnerabilities (which can evolve over decades) requires strategic planning, strategic investment, and strategic attention. To date, states and international institutions seem woefully unprepared for such strategic necessities. Moreover, environmental and human security, since they are contentious issues, often fall victim to the "do nothing" response because of their vulnerability-based conditions in which the clearly identifiable cause and the desired prevented effect are often ambiguous.

Some examples might help illustrate this claim. Although now well known that the United States, among so-called developed nations, has spent billions of dollars on *studying* the issue of environmental change—particularly regarding the issues of global warming and greenhouse gases—there still remain linkage problems between cause and effect. If the temperatures were to rise, by some estimates, between 1.4° C and 5.8° C over the course of the twenty-first century, there could be a concomitant rise in sea levels of between 9 and 88 centimeters.[22] Such a rise in sea level, although not of immediate concern to most nations, would be the single greatest national security issue for a nation such as the Maldives; in essence, such a sea level rise would mean the end of the Maldives (because the entire landmass would be under water).

A second example is equally striking. Because of rising temperatures—although no one precisely knows how high, at what rate, and how much the levels will fluctuate—Canada faces a unique conundrum. Sometime over the next 15 years, the Northwest Passage—which Canada claims as territorial waters—may become navigable. As such, a navigable Northwest Passage (which would cut the journey from Europe to Asia by 4,500 nautical miles, in comparison with transiting the Panama Canal) could lead to a rise in illicit crime, human trafficking, drug smuggling, pollution of the fragile Arctic ecosystem, human disasters at sea, and violations of Canadian sovereign territory.

Given the amount of study that has been devoted to such vulnerability-based security issues, there has been far less attention given to the potential strategic responses. The time, perhaps, for study has passed, and the time for action, implementation, and preparatory response may well have arrived. Daniel Esty phrases this dynamic well in describing how the notion of sustainable development is rapidly becoming a "buzzword largely devoid of content" and that new methods and ideas for action must quickly be set in place:

[The] world needs concrete pollution control and natural-resource management initiatives—for starters, a better global environmental regime, improved data and performance measurement and dissemination of environmental best practices, and a beyond-Kyoto climate change strategy.... The time for grand vision and flowery rhetoric has passed. The challenges ahead require sharper focus, real commitment, and concrete actions.[23]

In essence, we have moved from the dynamic of the traditional *security dilemma* to encompass issues in the twenty-first century that will include as well a new *human dilemma* in specific geographic locations that require sustainable development and long-term investment strategies. Plausible "creeping vulnerability" scenarios thus might reasonably include the following:

- Different levels of *population growth* in various regions, particularly between the "developed" and the "emerging" world—to incorporate disproportionate population growth—youth bulges—and unprecedented levels of urbanization unseen in human history;
- The outbreak and the rapid spread of *disease* among specific "target" populations (such as HIV/AIDS) as well as the spread of new strains of emerging contagions such as SARS and H5N1;
- Significant *climate change* due to increased temperatures, decline in precipitation, and rising sea levels;
- The *scarcity of water and other natural resources* in specific regions for drinking and irrigation, and the compounding growth among populations dependent on transboundary water resources;

- The *decline in food production, access, and availability,* and the need to increase imported goods;
- Progressing *soil erosion and desertification;*
- Increased *urbanization and pollution* in "megacities" (populations of 10 million or more) around the globe, with the recognition that in the Lagos-Cairo-Karachi-Jakarta arc from 2000 to 2020, most will migrate to urban environments that lack the infrastructure to support rapid, concentrated population growth;
- The need to develop *warning systems* for natural disasters and environmental impacts—from earthquakes to land erosion.

These emerging vulnerabilities will not mitigate or replace more traditional hard security dilemmas. Rather, we will see the continued reality of threat-based conditions contend with the rise of various vulnerability-based urgencies. Paradoxically, creeping vulnerabilities will likely receive the least attention, even as their interdependent complexities grow increasingly difficult to address over time.

A powerful illustration of urbanization shifts as creeping vulnerability is relevant to consider—again, a security potential especially critical when considered in its relationship to climate change and human impact. To be specific, truly cataclysmic demographic changes will occur in the Lagos-Cairo-Karachi-Jakarta arc in the coming decades,[24] where there will be astounding shifts in the global landscape that hinge on the flocking of populations to urban centers. According to the National Intelligence Council's *Global Trends 2015: A Dialogue about the Future with Non-Governmental Experts,* as well as from data compiled by the National Geographic Society in its November 2002 issue, and the United Nations Population Division (in the 2001 revision), world population in 2015 will be 7.2 billion and most will tend to live longer than they do today. Ninety-five percent of the increase will take place in emerging countries, and nearly all this population growth will happen in rapidly expanding urban areas.[25]

Consider, for example, the difference between growth in the emerging world versus growth in the so-called developed world from 1950 to projected figures for 2015: New York City will shift from 12 million inhabitants in 1950 to 17 million in 2015, whereas Lagos will grow from 1 million to 24.4 million; Los Angeles will shift from 4 million to 14.2 million, whereas Karachi will explode from 1.1 million to 20.6 million. The most stunning change, however, will occur in Dhaka, the capital of Bangladesh. From 1950 to 2015, the population is projected to grow from 400,000 to 19 million—a growth increase of 4,750 percent. Even minor reflection on this rapid shift reveals staggering change.

Urbanization of itself, of course, is value neutral—neither a good nor a bad thing. Urbanization, therefore, does not have to suddenly become a security issue or a threat. Tokyo, for example, will mushroom from 6.2 million to 28.7 in the period from 1950 to 2015, yet it will likely be far better equipped and able to handle the infrastructure requirements of the future mega-city—unlike many cities in the emerging world. And although Tokyo will remain the world's largest populated city, in many ways Japan has already accommodated for its urbanized existence. (At this writing, 72 percent of Japanese live in cities.)

It is unclear, however, whether Lagos, Dhaka, or Tehran can sustain growth rates such as those projected here. Indeed, it is not even clear if many cities in developed states could sustain such rates. Dhaka's projected growth rate from 1950 to 2015, compared with New York City's growth over the same time frame, suggests that, were New York City to grow at an equivalent rate, the Big Apple in 2015 would be the *really* Big Apple, with a population just shy of 600 million people—twice the current population of the United States. As it seems unlikely that advanced polities such as New York state would be able to sustain such rapid growth, how can we assume that Bangladesh—already one of the most impoverished states on the face of the earth—could possibly accommodate a dramatic surge in the population of its capital city? Added to this volatile mix, of course, is the reality is that climate change and environmental impact—most especially water distribution and severe monsoon flooding—are already severely affecting the people of Bangladesh and the region surrounding it.

The real effect of urbanization, and where it will most rapidly take place, also reveals itself in the projection for the year 2015, where the number of cities with population of more than 5 million will skyrocket from 8 (in 1950) to 58. In addition, various population studies suggest that it could be possible to see more than 600 cities worldwide with populations in excess of 1 million inhabitants by 2015; in 1950, by contrast, there were only 86 such cities on the planet.

Given the extended example of urbanization as a creeping vulnerability, it should not be difficult to grasp why the other vulnerabilities listed here—youth bulges, disease outbreak, climate change and human impact, resource scarcity, soil erosion and desertification—are interconnected. As one factor tends to distort beyond control, other factors tend to follow.

All of these factors lead to what Richard Norton of the U.S Naval War College has described as the "Feral City" syndrome: cities that have grown beyond the "natural" carrying capacity of their respective national and civic governments to provide sufficient security within "pockets of darkness" in their municipal boundaries.[26] Feral cities will exist within states, nonetheless, clearly linked to the globalization process—with commercial, communications, and transportation links to the rest of the world. Examples include the *favelas* in Rio de Janeiro and ungovernable areas in Johannesburg and Gauteng Province in South Africa.

Violent crime and sexual offenses, furthermore, now account for almost one-third of reported offenses in many urban environments in the emerging world. In Rio de Janeiro, murder rates reach 60 per 100,000 residents; in Calcutta, 91 per 100,000; and in Johannesburg, 115 per 100,000.[27] In Rio, where the infamous Brazilian *favelas* have long been sites for criminal control and where policing actions are unable to cope or control activity within these feral zones, many *favelados* feel marginalized, and live in a pervasive atmosphere of fear—as much afraid of police as of drug lords.[28]

In Karachi, the most violent and lawless city in Asia, 40 percent of the population inhabit *katchi abadis* (slums), a fertile base for radical Islamism, and the city itself (Pakistan's largest and its biggest seaport as well) is a conduit for arms smuggling to the outside world.

In Lagos, Nigeria, the city's population has already exploded. Lagos today suffers from high unemployment and massive youth bulges; it is the nucleus of constant turmoil. Indeed, fighting in Lagos between the Yoruba and ethnic Hausa is thought to have far more to do with poverty and lack of opportunity than ethnic hatred. Lagos is also the center of Africa's international criminal network, and pervasive crime and corruption have crippled the economy, contributed to social and political tensions, and undermined relations with major potential trading partners in North America and Europe.

A second example of creeping vulnerability, one in which outcomes do not *necessarily* lead to violence, can found by a complex interrelationship among water use, agriculture, and the expectations of emerging societies and adapting lifestyles of the future. To briefly offer an example of this complex—yet potentially serious—creeping vulnerability, consider that in 1900, the earth's population was 1.6 billion people; in 2000, that number reached 6 billion. In 1900, a male American had a life expectancy of 47 years; in 2000, that life expectancy reached 77 years. Notably, exploding water consumption from 1900 to 2000—from roughly 500 cubic kilometers to 5,000 cubic kilometers—was not directly linked to increased populations growth per se. Rather, the real "explosion" in water usage, where over the past three centuries water consumption has grown by a factor of 45, suggests that the real culprit is water usage for agriculture.[29] Seventy percent of all water use accounts for agricultural purposes. Yet the problem becomes even more complicated.

In 1960, roughly, all available arable land ran out. Since then, the world's population has doubled and grain yields—wheat, rice, corn—tripled. As an example of how a developed state abuses water resources for a developed lifestyle, consider these linkages: Eighty-two percent of American cropland—corn, unprocessed wheat, hay, soybeans—is not grown for consumption. Rather, these are grown for other food sources—refined and processed foods, and feed for livestock. Much of the emerging world is following America's example. In 1960, Mexico fed 5 percent of its grain to livestock; today, Mexico feeds 45 percent of its grain to livestock. Egypt went from 3 percent to 31 percent over the same time period. China, with a sixth of the world's population, has gone from 8 percent to 26 percent.[30]

This second example, in particular, points to how creeping vulnerabilities cannot—unlike threats—be couched exclusively in terms of negative outcomes. To the contrary, these vulnerabilities present opportunities, challenges, risks, and serious need to recognize and adapt to their increasing influences. Moreover, failing to recognize the resident strength or weaknesses in how states will continue to act in the future to new and old security challenges may well lead to increasing instances of "boomerang effect."

There is, indeed, no denying that there is a hazard in adding the term "security" to either environmental or human-centered concerns. That hazard is analytical imprecision and its practical consequences—using the wrong tools for the right missions and forsaking the opportunity to establish focused institutions that are designed, unlike the military, for the purpose of addressing what are essentially nonmilitary concerns. Modifying existing institutions may be helpful, but the reality remains that the solution—or at least the best response—to combating complex vulnerabilities lies in the nonmilitary realm. Making the armed forces more like a nonmilitary entity will not be enough to solve the nontraditional problems but will be enough to cause failure in addressing the traditional ones.

In such instances, the need for intervention is immediate, but as has been demonstrated in the past, the international community has tended to do little to prevent the ripples from forming and or rescuing Tawney's figurative "almost drowning" peasant.

RESPECTING RISK, UNDERSTANDING UNCERTAINTY

Risk is the term always mentioned in security discussions—usually mentioned last, usually in the context of "yes, it's important," and almost always ignored in the final decision process. Yet, in weighing the importance of risk, it seems increasingly important to recognize that it is the one factor that cannot be ignored.

At the height of the Cold War, for example, both stability and parity between the United States and the USSR was based on what could well be called a "balance of terror" with nuclear weapons. As such, this basic state of insecurity drove the international system *toward,* rather than away from, stability. This insecurity also influenced the recognition that *risk* was as much a driving force in the guarantees of basic security as the absence of fear or the desire to be free to make choices on behalf of the collective good. Risk, in our estimation, is also critical to understanding how—in an age of "new" security—humans will adapt to or suffer in response to the effects of climate change.

Yet as critical as it is to recognize the importance of acknowledging and dealing with risk, *risk* remains difficult to define in precise terms. (Similarly, in our earlier discussion introducing our concepts on the importance of vulnerability, we should also emphasize here that vulnerability scholars Vogel and O'Brien define *risk* in general terms as "the chance of a defined hazard occurring.")[31] Our own approach to defining—and by extension, understanding—risk involves an admittedly minimalist conception: *the ability to expose oneself to damage during the process of change and the resilience to be able to sustain oneself during such change.* Although many might well object to this identification as too broadly covering the category, we do find this approach useful in that our definition should remind one that risk, if respected and acknowledged, can never be assumed away. One useful example of how risk is dangerously assumed away is taken from contemporary East Africa. During a period when climate change has induced semipermanent drought effects in much of the region, natives have now come to depend, indeed expect, permanent food support through international agency distribution. Although this may seem an odd form of dependency to some, it can also incur fatal consequences when environmental impacts potentially expand beyond the region and food distribution would not, in the future, remain as assured as it is today.[32]

Ulrich Beck classifies risk in a highly provocative context—one related to human action:

Danger is what we face in epochs when threats can't be interpreted as resulting from human action. Rather than being experienced as decision-dependent, they are interpreted as being unleashed by natural catastrophe or as punishment from the gods. They are experienced as collective destiny. Risk, by contrast, marks the beginning of a civilization that seeks to make the unforeseeable consequences of its own decisions foreseeable, and to subdue their unwanted side effects through conscious preventative [*sic*] action and institutional arrangements.[33]

Related to risk, of course—and something that can never be separated from it—is uncertainty. Although slightly oxymoronic to claim that it is essential to "understand" uncertainty, it remains likely that it is essential to be able to appreciate its complexity and to consider its consequences and potential impact. Much of chapters 4 and 5 address the mind-set involved in approaching levels of uncertainty, and the difficult, though necessary, choices that scenarios offer in attempting to address climate change and human impact.

Admittedly, levels of uncertainty when considering the catastrophic risk of global climate change can prove maddeningly frustrating, as the graphics in figure 3.2, from the Union of Concerned Scientists, illustrate.[34]

Figure 3.2 shows that though there may be suggested leveling of temperature increase by the late twenty-first century, rapid increases in the early decades cannot allow for complacency. Figure 3.3 suggests a similar dynamic. The disturbing implication of this second graphic lies in continued sea-level rise decades after "heat-trapping gases are stabilized and the upward temperature trend levels off."[35]

Both graphics, of course, can prove maddeningly frustrating (for decision makers) in their variability. What proves "good science," by allowing for uncertainty, tends to go beyond the pale of allowable boundaries for many who might seek to implement actions to mitigate, adapt, or enhance resiliency for

Figure 3.2
Projected Temperature Increase

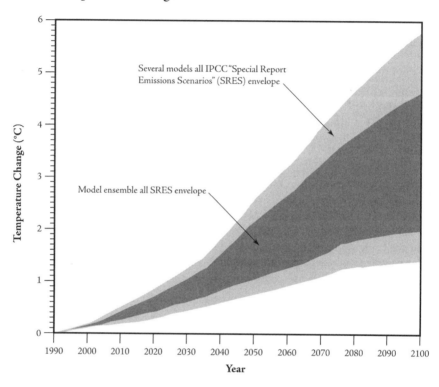

Temperature Change

those human victims who must suffer the consequence of such potentially immense change.

Although we examine how uncertainty affects mind-sets in greater consideration in chapter 5, it seems prudent to emphasize here how crucial risk and uncertainty are in the context of this book.

THE SHIFTING STRUCTURE OF SECURITY

To be clear, avoiding disastrous long-term impacts of creeping vulnerabilities such as the ones that will be generated by fast urban growth in weak states requires the innate and intuitive ability to look at the long view. (Indeed, Peter Schwartz once named this ability the "Art of the Long View.") To date, states and international institutions seem woefully unprepared for such strategic necessities. Moreover, environmental and human security, since they

Figure 3.3
Projected Sea-Level Rise

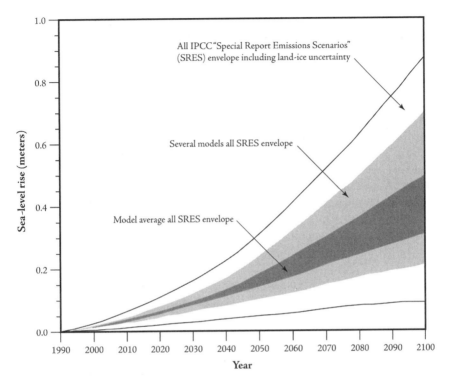

Sea-Level Rise

are contentious issues, often fall victim to the "do nothing" response because of their vulnerability-based conditions in which the causes are not clearly identifiable and are subject to debate. Another fact may be that political leaders have less confidence in the tools for remediation than in the traditional tools of military force and diplomacy at the state level, in particular military cooperation and diplomacy.

These examples underscore that no single instrument—no matter how seemingly powerful in its application—is sufficient to address new and emerging security issues. The old cliché that describes the "blowback" that might occur with the misapplication of strategic security tools (particularly the use of military force) is an apt reminder: *If all you have is a hammer, then every problem begins to look like a nail.* It might perhaps be more apt to say that when one only has a hammer, the problems that do not look like nails are ignored. An obvious solution would be to add new tools to the repertoire and thus equip ourselves to do something about the problems that are not nails.

Unfortunately, the obvious solution is one that governments have been re-luctant to choose because it would involve reallocating funds from existing programs with vested interests. Another reason may be the intuition that when these new tools are acquired, the result would be engagement in messy long-term projects that are not politically popular and, moreover, do not yield results that can be meaningfully measured on a quarterly or even annual basis. In de-mocracies, the ability to demonstrate results for overseas expenditures is crucial to the political success of the leaders responsible for the expenditures and for the continued funding of the overseas programs.

These emerging vulnerabilities will not mitigate or replace more traditional hard security dilemmas. Rather, we will see the continued reality of threat-based conditions contend with the rise of various vulnerability-based urgen-cies. By their very nature, creeping vulnerabilities will likely receive the least attention, even as their interdependent complexities grow increasingly difficult to address over time.

These developments and trends argue that it is time to rethink both the structure and the nature of our understanding of security. For all the talk in policy centers and the academy about new world orders, networks of interna-tional dependence, and the shrinking of the nation-state's power, the traditional definition of security has not been seriously challenged in the context of the trends that are going to shape the world in the next few decades. Climate change, as the catalyst for the acceleration of negative effects, could well prove to be the volatile fuel added to the already raging fire.

In terms of exact categorization, there are admittedly critical differences between human and environmental security. In the broadest sense, environ-mental security considers issues of environmental degradation, deprivation, and resource scarcity; by contrast, human security examines the impact of systems and processes on the individual, while recognizing basic concerns for human life and valuing human dignity. Yet as numerous examples illustrate, complex interactions within various environments often place stress on the security of the individual. Thus, environmental and human security often coexist in a compli-cated interdependence best conceptually considered as "extended security."

Policy makers would be wise to recognize this conceptual approach.[36] For research to be relevant to policy makers, it should almost always contextual-ize significance within a specific human- and regional-oriented perspective. (Though nonspecific in its references to human security, the Schwartz and Randall scenario approach [Appendix One], discussed at length in chapter 1, nonetheless attempts to describe both human impacts and possible regional security effects.) Yet Ole Wæver, one of the earlier influences (along with Barry Buzan) in promulgating the "new security agenda," reflects a certain skepticism, nonetheless, about the ability to influence policy through reframing the under-standing of security:

A security issue demands urgent treatment: it is treated in terms of threat/defence, where the threat is external to ourselves and the defence often a technical fix . . . traditionally the state gets a strong say when something is about security. To turn new issues (such as the environment) into "security" issues might therefore mean a short time gain of attention, but comes at a long-term price of less democracy, more technocracy, more state and a metaphorical militarisation of issues. For this reason, environmental activists and not least environmental intellectuals who originally were attracted to the idea of "environmental security" have largely stepped down. . . . Security is about survival. . . . The invocation of security has been the key to legitimizing the use of force, and more generally opening the way for the state to mobilize or to take special force. . . . Security is the move that takes politics beyond the established rules of the game."[37]

There are, however, any number of overextending assumptions in the preceding reference. Above all is the assumption that security is an extreme term that can only be couched in terms of threat, and that the state—as political monolith—can only respond to with the use of force. Security is far richer in contextual meaning than such a stratified identity seems to allow.

Security is a basis for both policy *response* and long-term *planning*. Furthermore, the use of force—particularly military force—is often an ineffectual and irrelevant response to the "new security agenda." Thus, the argument that "environmental security" is simply a mask for military intervention is at best thin. What *is* true is that the understandings of, and definitions for, environmental security range so broadly that its meaning takes on something for everyone—and perhaps, ultimately, nothing for any one.

In the broadest sense, environmental security centers on a focus that seeks the best effective response to changing environmental conditions that have the potential to reduce stability, affect peaceful relationships, and—if left unchecked—contribute to the outbreak of conflict. Although it seems attractive to insist on exclusionary concepts that insist on desecuritization, privileged referent objects, and the "belief" that threats and vulnerabilities are little more than social constructions, all these concepts work in theory but fail in practice. Though true that traditional and national security paradigms can, and likely will continue to, dominate issues that involve human security vulnerabilities—and even in some instances mistakenly confuse vulnerabilities as threats—there are distinct linkages between these security concepts and applications. With regard to environmental security, for example, Myers recognized these linkages decades ago:

National security is not just about fighting forces and weaponry. It relates to watersheds, croplands, forests, genetic resources, climate and other factors that rarely figure in the minds of military experts and political leaders, but increasingly deserve, in their collectivity, to rank alongside military approaches as crucial in a nation's security.[38]

Simply acknowledging the immensities of these challenges is obviously not enough.

Yet the creation of a sense of urgency to act—*even on some issues that may not have some impact for years or even decades to come*—is perhaps the only appropriate first response. The real cost of not investing, in the right way and early enough, in the places where trends and effects are accelerating in the wrong direction is likely to be decades and decades of economic and political frustration, and, potentially, military engagement. One approach, which might avoid the massive tidal impact of creeping vulnerabilities, is to sharply make a rudder shift from constant crisis intervention toward strategic planning, strategic investment, and strategic attention. Clearly, the time is now to reorder our entire approach to how we address—or fail to address—security.

We live in an age of zombie concepts and boomerang effects—what sociologist Ulrich Beck terms the "Global Risk Society." Undeniably, Beck suggests that with vast leaps in technological advance and interlinked communications, we have only increased the potential for local, and localized, risk to now mutate into *systemic risk*.[39] Drawing on concepts brought forth by John Maynard Keynes, systemic risk both invokes and involves radical uncertainty. The crucial error, therefore, is not to believe that there is simply data or information missing that could fill the uncertainty gap; to the contrary, "manufactured uncertainties," as Anthony Giddens terms this paradox, occur when increased knowledge fuels the production of ever more increased uncertainty.[40]

Regarding the science of climate change, the implications are astonishing. We do not suffer from insufficient information; we suffer from too much scientifically conclusive data. We cannot reduce our uncertainty in order to resolve the definitive debate on climate change and human impact; we can only increase our uncertainty with the insecurity of our increased knowledge.

Ultimately, we are no closer to addressing how best to solve these challenges, even as they collide with environmental, human, and national issues. As subsequent chapters argue, climate change will be the driving force and critical uncertainty that will alter everything.

NOTES

1. John Urry, "Introduction: Thinking Society Anew," in *Conversations with Ulrich Beck,* ed. Ulrich Beck and Johannes Willms, trans. Michael Pollak (London: Polity, 2004), p. 6.

2. Urry, pp. 7–8; Ulrich Beck, *World Risk Society* (London: Polity, 1999), p. 116.

3. Elizabeth Kolbert, *Field Notes from a Catastrophe: Man, Nature, and Climate Change* (London: Bloomsbury, 2006), p. 187.

4. There is considerable debate, as highlighted in this book, as to how far this conceptualization should go. Indeed, there is a degree of disagreement between the authors, discussed subsequently. Whereas Owen believes that human security should be limited to the very worst threats in any one location, Liotta sees it as a more holistic concept encompassing wider vulnerabilities and the mechanisms to counter them.

5. *The Leviathan,* edited and with an introduction by C. B. MacPherson (London: Penguin Books, 1985).

6. United Nations Development Programme (UNDP), *Human Development Report: Annual Report* (New York: UNDP, 1994), pp. 22–23.

7. United Nations Commission on Human Security, *Protecting and Empowering People* (New York: United Nations Publications, 2003), http:www.humansecurity-chs .org/finalreport/outline.html.

8. UNDP, "New Dimensions of Human Security," in *Human Development Report: Annual Report* (New York: UNDP), pp. 22–25.

9. Emma Rothschild, "What Is Security? The Quest for World Order," *Dædulus: The Journal of the American Academy of Arts and Sciences,* 124 (June 1995): p. 51.

10. Earlier versions of this still-evolving table were partially inspired by Bjorn Møller, "Global, National, Societal, and Human Security: A General Discussion with a Case Study from the Middle East," presented at the 4th Pan-European International Relations Conference in Canterbury, United Kingdom, and as part of Bjorn Møller, "National, Societal and Human Security: Discussion—Case Study of the Israel-Palestine Conflict," in *Security and Environment in the Mediterranean: Conceptualizing Security and Environmental Conflict,* ed. Hans Günter Brauch, P. H. Liotta, Antonio Marquina, Paul Rogers, and Mohammed El-Sayed Selim (Berlin: Springer Books, 2003), as well as work presented to P. H. Liotta, the Faculty of Civil Defence in Belgrade, Serbia, and published as "National, Societal and Human Security," *Ljudska Bezbednost* [Human Security] 1:1 (2003). Note that the table shown here is markedly different in several critical areas. The author fundamentally argues, through the use of the boomerang effect example, that those in the so-called Copenhagen School—who have focused on the terms "referent object," "societal security," and "desecuritization"—are inevitably delinking their arguments from effective use in the foreseeable future for policy decisions of most states.

11. Jen Bissell, "A Comet's Tale: On the Science of Apocalypse," *Harper's* (February 2003): 35.

12. A sufficient review of the literature of human security is not possible in the available space here. Some of the most stimulating pieces on the subject of environmental and human security are found in *Security and Environment in the Mediterranean: Conceptualizing Security and Environmental Conflict,* ed. Hans Günter Brauch, P. H. Liotta, Antonio Marquina, Paul Rogers, and Mohammed El-Sayed Selim (Berlin: Springer Books, 2003): Bjorn Møller, "National, Societal and Human Security: Discussion—Case Study of the Israel-Palestine Conflict"; Nils-Petter Gleditsch, chapter 26, "Environmental Conflict: Neomalthusians vs. Cornucopians." Other work includes Jorge Nef, *Human Security and Mutual Vulnerability: The Global Economy of Development and Underdevelopment,* 2nd ed. (Ottawa: International Development Research Centre, 1999); Roland Paris, "Human Security: Paradigm Shift or Hot Air?" *International Security* 26 (Fall 2001), pp. 87–102; Peter Stoett, *Human and Global Security: An Explanation of Terms* (Toronto: University of Toronto Press, 1999); Caroline Thomas and Peter Wilkin, eds., *Globalization, Human Security, and the African Experience* (Boulder, CO: Lynne Rienner, 1999); Joseph Stiglitz, *Globalization and Its Discontents* (New York: W. W. Norton & Co., 2002); Majid Tehranian, ed., *Worlds Apart: Human Security and Global Governance* (London: I.B. Tauris, 1999); Tatsuro Matsumae and L. C. Chen, eds., *Common Security in Asia: New Concept of Human Security* (Tokyo: Tokai University Press, 1995); Yuen Foong Khong, "Human Security: A Shot-

gun Approach to Alleviating Human Misery?" *Global Governance* 7 (July-September 2001). Moreover, a recent issue of *Security Dialogue* 35(3): 345–87, displays the rich diversity and division of perspectives on "What Is 'Human Security'?" Taylor Owen's concluding essay, "Human Security—Conflict, Critique, and Consensus: Colloquium Remarks and a Proposal for a Threshold-Based Definition" (pp. 373–87), is especially useful.

13. UN Commission on Human Security, *Human Security Now* (New York: United Nations Publications, 2003), p. 4.

14. For one perspective that suggested that a convergence between human and national security was occurring in the period after September 11, 2001, see P. H. Liotta, "Boomerang Effect: The Convergence of National and Human Security," *Security Dialogue* 33:4 (2002): pp. 473–88, as well as the responses of: Brooke A. Smith-Windsor, "Terrorism, Individual Security, and the Role of the Military: A Reply to Liotta," *Security Dialogue* 33: 4 (2002): pp. 489–94; P. H. Liotta, "Converging Interests and Agendas: The Boomerang Returns," *Security Dialogue,* 33: 4 (2002): pp. 495–98.

15. Accordingly, we witness 150,000 American forces deployed to Iraq in 2003 but only 6 communication command and control specialists initially put ashore in Liberia to support Nigerian peacekeepers in stability and security operations. Nominally, both scenarios involve regime change and stabilizing regional security—as well as intervention for the protection of citizens abused by the state. Yet the physical and economic geography of Iraq place it at the center of a region declared "vital" to U.S. interests, awash in petroleum and natural gas resources. National security interests, in the form of geopolitics, again trumps the intervention priority list. Moreover, the alleged bellicosity of the former Iraqi regime, particularly regarding potential or actual possession of weapons of mass destruction, clearly supported more traditional national security interests such as defense of the homeland and protection of one's territory from attack. Liberia, though clearly a regional troublemaker, never posed a "threat" to the United States or any of its close allies. (Make no mistake: It was a brutal, authoritarian regime that threatened its own people as well as the entire security architecture of west Africa but remained little more than a perceived peripheral threat for many.)

16. George Packer, "War and Ideas," *The New Yorker,* July 5, 2004, p. 32.

17. Drawn from an abstract of a presentation at the Pell Center for International Relations and Public Policy, Newport, Rhode Island, titled "Culture Wars? War Is Already a Culture," December 6, 2004, at a workshop titled "Prepared for Peace? The Use and Abuse of 'Culture' in Military Simulations, Training and Education." Professor Burgess refers to Mary Kaldor, *New and Old Wars: Organized Violence in a Global Era* (Palo Alto, CA: Stanford University Press, 1999).

18. Hans-Georg Bohle, "Vulnerability Article 1. Vulnerability and Criticality," *IHDP Update: The Newsletter of the International Human Dimensions Programme on Global Environmental Change* 2 (2001). http://www.ihdp.uni-bonn.de/html/publications/update/update01_02/IHDPUpdate01_02_bohle.html. Vulnerability aspects are adapted from Coleen Vogel and Karen O'Brien, "Vulnerability and Global Environmental Change: Rhetoric and Reality," *Aviso: An Information Bulletin on Global Environmental Change and Human Security,* 13 (March 2004), p. 6.

19. James C. Scott, *The Moral Economy of the Peasant: Rebellion and Subsistence in Southeast Asia* (New Haven, CT: Yale University Press, 1976), pp. iv, 1.

20. P. H. Liotta, "Chaos as Strategy," *Parameters,* Summer 2002, pp. 47–56.

21. Hugh Courtney, Jane Kirkland, and Patrick Viguerie, "Strategy under Uncertainty," *Harvard Business Review* 1 (November 1997): pp. 66–79.

22. These data are based on the Summary for Policymakers of the Intergovernmental Panel on Climate Change (IPCC) that were approved in January 2001 in Shanghai by the IPCC member governments. As such, they do not offer definitive, discrete "proof." Final, convincing, and irrefutable data for these issues do not exist. Equally, the estimate of 5.8° Celsius exceeds the estimates of recent UN and the American National Academy of Sciences data. These illustrations are meant to show the nature of security issues that rise out of vulnerabilities rather than out of direct threats. The issues themselves and the best responses to these issues lack the precision and clarity of threats. IPCC, *Climate Change 2001: The Scientific Basis* (Cambridge, UK: Cambridge University Press, 2001), pp. 13–14.

23. Daniel C. Esty, "A Term's Limits," *Foreign Policy* (September–October 2000): 75.

24. For specific explanation of the "Lagos-Cairo-Karachi-Jakarta arc," see James F. Miskel and P. H. Liotta, *A Fevered Crescent: Security and Insecurity in the Greater Near East* (Gainesville: University Press of Florida, 2006).

25. Central Intelligence Agency, *Global Trends 2015: A Dialogue about the Future with Nongovernment Experts* (Washington, DC: 2001), http://www.cia.gov/cia/publications/globaltrends2015/index.html.

26. Richard J. Norton, "Feral Cities," *Naval War College Review* 56, no. 4 (Autumn 2003): pp. 97–106.

27. United States Government Interagency Working Group, *International Crime Threat Assessment* (December 2000). http://www.fas.org/irp/threat/pub45270index.html; Mayra Buvinic and Andrew R. Morrison, "Living in a More Violent World" *Foreign Policy* 118 (2000): pp. 58–72; United Nations Centre for Human Settlements, *Global Report on Human Settlements* (Oxford, UK: Oxford University Press; United Nations Commission on Human Security, 2003. *Protecting and Empowering People,* http:www.humansecurity-chs.org/finalreport/outline.html; United Nations Development Programme (UNDP), *Human Development Report: Annual Report.* New York: UNDP, 1994.

28. Janice Perlman, "The Metamorphosis of Marginality: From Myth to Reality in the Favelas in Rio de Janeiro, 1969–2002" (unpublished draft, 2002), quoted in Ellen Brennan-Galvin, "Crime and Violence in an Urbanizing World," *Journal of International Affairs* 561 (2002): p. 123.

29. Fen Montaigne, "Water Pressure" *National Geographic,* September 2002, pp. 2–33, 16.

30. Richard Manning, "The Oil We Eat: Following the Food Chain Back to Iraq," *Harper's,* February 2004, pp. 37–45.

31. Vogel and O'Brien, 5–6.

32. Drawn from a joint lecture by Patricia Kameri-Mbote, Africa Policy Scholar, Woodrow Wilson International Center for Scholars, and Program Director, International Environmental Law Research Centre, University of Nairobi, Kenya; and Dr. Geoffrey Dabelko, Director, Environmental Change and Security Program & Coordinator, Global Health Initiative, Woodrow Wilson International Center for Scholars, Washington, DC, "Water Woes: The Critical Importance of Scarce Resources," presented at the Pell Center for International Relations and Public Policy, Newport, Rhode Island, April 24, 2006.

33. Beck and Willms, "Global Risk Society," *Conversations with Ulrich Beck,* pp. 110–11.

34. http://www.ucsusa.org/global_warming/science/projections-of-climate-change.html.

35. Ibid.

36. In furthering this argument, this chapter intentionally blurs the distinction between environmental and human security, in favor of an extended security approach. For the best overall writing and conceptualization of "extended security," reference the work of Emma Rothschild, Director of the Centre for History and Economics, Kings College, Cambridge University. As difficult as such "new" conceptions of security are, it seems worth noting that the United Nations Commission on Human Security continued to grapple with the definition of *security.*

37. Quoted in Thomas Scheetz, "The Limits to 'Environmental Security' as a Role for the Armed Forces," unpublished paper provided by the author. Wæver's original remarks, titled "Security Agendas Old and New, and How to Survive Them," were prepared for the workshop on "The Traditional and New Security Agenda: Influence for the Third World," Universidad Torcuato Di Tella, Buenos Aires, Argentina, September 11–12, 2000.

38. Norman Myers, "The Environmental Dimension to Security Issues," *The Environmentalist* (1986): 251.

39. Beck and Willms, "Global Risk Society," *Conversations with Ulrich Beck,* p. 127.

40. Ibid.

— 4 —
Through the Glass Darkly: Scenarios and Alternative Futures

The true voyage of discovery lies not so much in seeking new landscapes as in having new eyes.

Marcel Proust

Several years ago, as one of the authors was making a formal presentation on strategic planning to the country team of the American embassy in Madrid, the senior political counselor suddenly burst out, "Why should anyone *care* about strategy? It's hard enough dealing with *policy,* going from one crisis to the next!" To be fair, for this foreign service officer, who had recently experienced any number of policy crises—from Haiti to the Balkans—there was a point to his objection. Why *should* anyone care about strategy?

Strategy, after all, is not politically expedient; it is a long-term focusing instrument that helps shape the future environment.[1] Policy crises, on the other hand, always deal with the more immediate execution of initiatives to address critical needs and requirements. But if an argument could be made in defense of strategy, it would be this: In the absence of strategy, there is no clear direction for the future, and any road will take you there, bumping over crisis and change, and suffering through one knee-jerk reaction after another.

Perhaps what best illustrates this reality is the scene between Alice and the Cheshire Cat in *Alice's Adventures in Wonderland,* when Alice asks, "Would you tell me, please, which way I ought to go from here?"

"That depends a good deal on where you want to get to," said the Cat.

"I don't much care where—" said Alice.

"Then it doesn't matter which way you go," said the Cat.

"—so long as I get *somewhere,*" Alice added as an explanation.

"Oh, you're sure to do that," said the Cat, "if you only walk long enough."

At its best, strategy will get you somewhere near where you intended to go. Strategy provides a systematic approach to dealing with change, with both what should and should not be expected to remain the same. Strategy, in short, is the application of available means to secure desired ends. As regards policy decisions, nonetheless, the outburst of the foreign service officer seems all too often the likely response to both crises and longer-term developing security conditions than simply immediate intervention choices. The issue of climate change and human impact, in particular, requires, we argue, a more focused, more nuanced, more strategic approach.

We therefore make a case for scenarios as useful strategic tools in allowing policy and decision makers to assess potential outcomes long before (sometimes lethal) impacts occur. Well-known scenario cases that have considered climate change and its security impact (e.g., the Schwartz and Randall case written for the U.S. Department of Defense's Office of Net Assessment)[2] have most often focused on how to mitigate risk, particularly in the context of "national security." Notably, such scenarios focus on the threat to security inherent in rapid environmental change, and few scenarios—to date—have addressed the possibility of how best to mitigate risk, lessen vulnerability, and possibly even limit threats.

APPROACHING SCENARIO-BASED STUDIES: WAYS TO REPERCEIVE THE FUTURE

Scenario-based investigations of possible futures have been used since the middle of the twentieth century to help decision makers cope with alternative courses of action and elements of uncertainty. Each scenario-based study is founded on assumptions of possible change. Scenarios begin with the intuitive sense that the ultimate success of decisions made today rests on the situation tomorrow. Unfortunately, despite the importance of the eventual future for current plans, our knowledge of it is usually, and at best, precarious. Indeed, it may not be an overstatement to say that we have precious little *knowledge* of the future at all.

Because the future has not happened, it offers no facts, presents no testimonies, and provides no means for immediate verification. Instead, there are only assumptions of the future: assumptions about the people, places, and things that might constitute the world of tomorrow; assumptions about how these constituents might interact; and assumptions about the impacts that might transpire.

An approach to help manage the inherent uncertainties of decisions based on assumptions, rather than on facts, is to examine several alternatives of how the future might unfold and compare the potential consequences of different future contexts. By doing so, a person, organization, company, or government can make decisions that are more resilient to the throes of tomorrow. In general, these alternative views of the future are referred to as *scenarios*. Dictionaries typically define the word *scenario* as a summary of the events of a play, film, or novel. But in the 1950s Herman Kahn appropriated the term for long-range visions of the future,[3] recasting the scenario as a "hypothetical sequence of events constructed for the purpose of focusing attention on causal processes and decision points."[4] Since then, many have contributed to the development of techniques for crafting and using scenarios.[5] In turn, some have added nuances to the term, often to emphasize a given methodological aspect, such as quantification, qualification, a level of precision or generalization, who participates and how, or the inclusion of past trends.[6] Most conceptions of scenarios share four principles:

1. scenarios are fictional (where *fictional* is understood to mean unverifiable but plausible, not fanciful) accounts that represent a process of change over some duration;
2. scenarios describe situations, actions, and consequences that are contingently related;
3. scenarios are understood to be *predictive judgments* that describe what could happen, not *predictions* that describe what will happen, or even what is likely to happen; and
4. scenarios organize information within explicitly defined frameworks.

As Kahn and others have argued, there are several benefits to using scenario-based approaches as part of a decision-making process. First, scenarios provide an aid, or more formally a heuristic, for defining conditions and assessing consequences. Facts and data are, in themselves, largely meaningless unless they are placed within an intellectual and disciplinary context. Scenarios provide a means to relate and comprehend isolated pieces of information within a single framework and—equally important—a structured approach by which to consider individual factors across different frameworks. Second, the specificity of description required in scenario creation—the who, what, where, when, and why of actions—can lead a decision maker to consider implications that would have been missed had the representation of the subject matter been limited to abstract principles. That is to say, the specifics of how change occurs may matter as much as, or even more than, the resulting situation. Third, because scenarios are fictional, they can serve as artificial case studies that illustrate implications of policies otherwise ignored or missed if examples from the past "real" world were considered exclusively. To the degree that the future will offer unprecedented situations, basing decisions only on the evidence of the past can be, at best, shortsightedly careless and, at worst, recklessly irresponsible.

Finally, scenarios provide a means to facilitate the discussion of planning options across stakeholder groups, professional disciplines, and levels of management. Different levels of government and different agencies within a level of government operate in unique jurisdictions, respond to different needs, and act with different means. Providing a single platform on which to discuss localized implications of the future can allow for a coordinated response to challenges.

SCENARIOS AND ALTERNATIVE FUTURES

As a primer to discussing the multiple facets of one's perception of the future, we offer here what are the basic principles of a scenario-based study. Figure 4.1 provides a graphic representation of several important parameters. The circle on the center grid represents a mapping of the present conditions relative to two concerns, one charted on the x axis and one charted on the y axis. The plotted issue might be readily pinpointed in quantitative terms such as the concentration of carbon dioxide in the atmosphere or be a perception of qualitative mapping such as the degree of economic conservatism held by a person or party in political power. Of course, reducing the understanding of the world to two issues is a simplification; however, representing the infinite number of interests that comprise modern society on a two-dimensional page is complicated, if not confusing, and so for graphic convenience only two are used to illustrate the concept.

Piercing the center of the grid, a double-headed arrow represents time. To the left is the past, along which are other grids that mark specific moments of history. Again, circles represent the conditions of those times mapped relative to the two axes of concern. Connecting the circles are solid lines representing the sequence of events that led from one period to the next. Multiple lines reflect the situation that the evolution from one state to the next may be explained in different ways, depending on one's point of view and the emphasis given to different influences. Some understanding may be found in each of the alternative explanations; moreover, the advancement of historical knowledge will undoubtedly add still more perspectives.

To the right of the center grid lies the future. Again, other grids mark specific times, on which are multiple dots representing alternative future conditions. The dashed lines that connect periods of time are the scenarios. Like histories, scenarios explain how the world could move from one state to another; however, as noted previously, unlike histories, which are based on verifiable facts, scenarios are based on assumptions.

In the literature on the use of scenarios, the terms *scenario* and *alternative future* are sometimes used interchangeably. There is, however, considerable methodological benefit in differentiating the expressions: an alternative future being defined as a possible end state; and a scenario being defined as a means to achieve that state.

Figure 4.1
Scenarios and Alternative Futures. Source: Allan W. Shearer, "Approaching Scenario-Based Studies: Three perceptions About the Future and Consideration for Landscape Planning." *Environment and Planning B: Planning and Design–Pion Limited, London,* **32,** no. 1 (2005), p. 71.

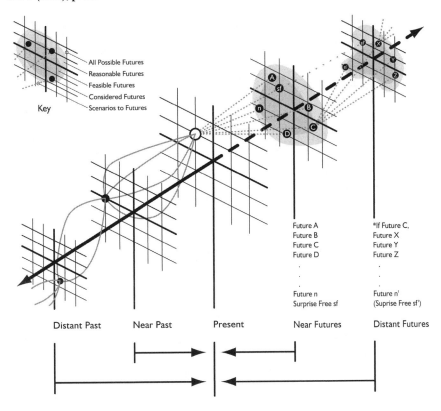

As Kahn notes, for example, a given future state may be achieved by more than one set of actions, and so, just as there can be several historical explanations of the past, there can be different scenarios leading to the same future.[7] Although the diagram shows only a single scenario line connecting the present to each alternative future, several lines could exist. A second reason to distinguish the terms can be understood through an analogy to motion pictures. A reel of film goes through a movie projector at 24 frames per second—more than fast enough to provide the spectator with a perception of movement, but too fast for him or her to make full sense of a single frame. A scenario is like a projected film in that it allows a representation of the dynamics that move us through different situations. An alternative future is like a single frame in that it gives us a snapshot, or cross-section, of time. From this stopped frame, we can carefully judge relative positions, take measurements, and assess impacts. Both the moving film strip and the stopped single frame are helpful for understanding the possible future and its implications; however, they are not valuable in the same way, and we benefit by having both. In addition, not all studies give equal emphasis both to the events leading up to a future and to the future itself. Some, for example, focus solely on the end state and leave the means necessary to achieve it for others to consider. As a result of these concerns, preserving the distinction between scenarios and alternative futures can help to differentiate kinds of investigations.

Of special note is the "surprise-free" alternative future, which is marked "sf" on the diagram. Sometimes referred to as "the trend" or "business as usual" expectation, it is the future that can be anticipated if there are no significant changes in the social, political, economic, technical, or environmental aspects of the world. In part, such stability precludes changes *in kind,* such as the development of a new, widely dispersed technology.

Development of the surprise-free future early in an investigation can have several benefits. Foremost, it captures a representation of the conventional thinking or the mental models by which people understand the world. Just as each person's or organization's point of view on a given issue can be understood by the interrelated aspects of ideology, stakeholder status, and physical position,[8] so mental models of how situations and actions interrelate are also unique. Articulating a surprise-free scenario provides a platform on which such thoughts can be openly examined and through which underlying assumptions can be identified. At extremes, it may be used as a straw man, against which preferable or better defined options might be considered. More neutrally, the surprise-free scenario can be beneficial as a reference point for people or groups who are not accustomed to formal scenario techniques.[9] Regardless of when in the process it is created, the surprise-free future can serve as a useful baseline against which the other alternatives can be compared.

Within each grid of the future are shaded areas that classify the alternatives as *possible, reasonable,* and *feasible.* A common concern of scenario-based studies is the need to seek visions that are novel and surprising while avoiding futures that are so far removed from reality that they are perceived as fantasy and summarily deemed irrelevant to the decisions at hand. The terms given here can serve as general qualifications for viability but are intentionally relative and purposefully subject to debate. It may sometimes be useful to assign probabilities to the likelihood that a given future will materialize; however, doing so can project a false precision—again, the future has not yet happened, and the situation tomorrow may have no precedent on which to base probabilities. Moreover, it has been argued that, because scenarios represent *possibilities,* all should be given equal consideration.[10] Assigning fixed probabilities can result in some alternatives being erroneously discounted and others inappropriately favored.

Distinguishing among possible, reasonable, and feasible is a pragmatic compromise between the need to assess the potential viability of a future and the desire to avoid erroneous claims of precision. Discussing scenarios in these terms can also be useful for gaining a better understanding of an investigation and for identifying possible limitations. A study that examines only alternative futures that are at the outer bounds of credibility may be beneficial for understanding extreme dimensions of risk but may also underestimate the opportunities and liabilities of futures that may be more readily realized. Similarly, a study focused only on highly plausible futures may miss novel but readily imaginable conditions.

We should also note that the assessment of possible, reasonable, and feasible is contingent on changing conditions. From the perspective of the present, the surprise-free scenario may seem to be a feasible option for a considerable amount of time. But, as the diagram shows, should alternative C occur in the near future, then moving back to that previously defined as surprise free in the distant future may be more complicated.

In figure 4.1, the more removed a time period is from the present, the smaller it is drawn. This convention reflects the broad presumption that the more distant an era is from one's own, the more difficult it can often be to imagine and understand its consequences. However, this generalization should not be mistaken for a universal theorem that all aspects of the future become equally more difficult to anticipate as the time horizon is extended. Looking backward at history, some parts of the distant past are known better than more recent events. We know more about the Roman Empire circa A.D. 100, for example, than we do about Native American settlements circa A.D. 1000. Similarly, assumptions about some aspects of the distant future may be more apparent and taken with greater confidence than others of the near future.

One final consideration here: The authors readily admit that they are of two minds regarding the omission of alternatives that may, at first glance, lack credible boundaries. On the one hand, Shearer, with a deep background in the pragmatic application of scenarios and alternative futures to landscape and environmental design problems, intentionally limits these credibility considerations. At the same time, he fully acknowledges that lack of consideration for highly *implausible* futures may prove problematic when applying scenarios and alternative future discriminative states to the complex, dynamic, and highly interactive conditions that involve climate change and human impact. On the other hand, Liotta, with a background in policy analysis and military force structure planning, argues that well planned and equipped forces are almost always forced to change, not by strategic design, but by the impact of the "unknown" that changes the rules of the game. Witness, for example, the surprise of the 1941 attack on Pearl Harbor, or the events of September 11, 2001—which some have described as the Pearl Harbor assault on liberal democracies by an ideology that is irreconcilable with the basic tenets of all admittedly diverse liberal democracies.

The cliché that describes this state of unknown events (and actions) that force change is well known, of course: Thomas J. Kuhn described it as a "paradigm shift."[11] There are, nonetheless, limits to how much consideration and how much planning should be devoted to planning for the unknown. There exist, after all, limits to planning for the impractical rather than for the pragmatic, and for outcomes that can be addressed or even solved in advance rather than for outcomes that cannot be solved, or perhaps even adapted to. Equally, one is further limited by the allocation of available resources. The U.S. military, for example, with more than half a trillion dollars in annual defense budgets (representing a budget that exceeds all of the rest of the world's defense budgets *combined*), is itself most restrained in its resource allocation choices. Difficult as it may seem to believe, there are not enough resources for all the problem set considerations one would like to address. Even the U.S. military, therefore, is limited by "scarcity" in its choices and options.

All this said, nonetheless, the authors can agree that—as regards climate change and human impact—there may emerge a deep necessity to consider the unimaginable, to think, in Kahn's terms, "the unthinkable." This issue boils down to how to deal with impact and the consequence of catastrophe. (Indeed, despite all the inflated press reports and the misinterpretations of the report itself, the implausible catastrophe recognition of Schwartz and Randall's "abrupt climate change" scenario offered a potential, and potentially necessary, way to consider the disastrous effects of climate change and human impact.) One of the best ways to consider such catastrophe lies in what is termed the "strangelet scenario."[12]

Sir Martin Rees, professor at the University of Cambridge and the United Kingdom's Astronomer Royal, hypothesized in the late twentieth century that, given our limited knowledge of subatomic particles, we could not, or should not, rule out the possibility of a doomsday scenario under which, in a desire to learn more about the largely unknown, a powerful particle accelerator might produce a shower of quarks that would

reassemble themselves into a very compressed object called a strangelet.... A strangelet could, by contagion, convert anything else it encountered into a strange new form of matter....A hypothetical strangelet disaster could transform the entire planet Earth into an inert hyperdense sphere about one hundred metres across.[13]

Although Rees admits to the high improbability of such a scenario, he also posits that with even a probability of one in a billion, such an outcome is not outside the boundaries of consideration—especially when the consequence of such a disaster would involve the complete elimination of the planet itself.

Although there are some complete dismissals to this scenario, one response to it was quite intriguing: The director of the Brookhaven National Laboratory (whose Relativistic Heavy Ion Collider [RHIC] was the focus of Rees's original thoughts on the strangelet scenario) commissioned a risk assessment by a committee of physicists before RHIC began operations. And, though the RHIC did eventually begin operations, the director of the assessment, John Marburger, offered that

All particles ever observed to "contain" strange quarks have been found to be unstable, but it is conceivable that under some circumstances stable strangelets could exist. If such a particle were also negatively charged, it would be captured by an ordinary nucleus as if it were a heavy electron. Being heavier, it would move closer to the nucleus than an electron and eventually fuse with the nucleus . . . ending up in as a larger strangelet. If the new strangelet were negatively charged, the process could go on forever.[14]

Added to the complexity of this condition is the reality that more powerful accelerators than the RHIC are on line today. Fermilab's Tevatron (Illinois) accelerator collides individual protons and antiprotons, generating energies greater than those at RHIC. The Center for European Nuclear Research in Geneva will begin operations in 2007 that will outdo RHIC in "luminosity"—roughly, how likely it is for collisions to occur in an accelerator. Finally, RHIC itself has sought funding for a "RHIC-II" upgrade, to begin operations in 2010, that would allow particle collisions up to 40 times the luminosity of the original RHIC accelerator.[15]

Although it is unlikely that a total disaster might occur from the results of these new scientific procedures, one can never—or should never—discount the possibility of its occurrence. As Richard Posner, author of *Catastrophe: Risk and Response* suggests, "one wouldn't want to bet the planet on it."[16]

ATTITUDES TOWARD THE FUTURE

In addition to the objective morphology of a scenario-based study, one must also consider the objective needs by stakeholders for the undertaking. The first perception about the future is one's position regarding the prospect of change. Russell Ackoff identifies four such attitudes.[17] As a preface, all of these attitudes can (and do) coexist, and a given person's or organization's attitude may evolve not only with times, but also with the specific aspects of change that is under consideration. The *reactive* attitude is typified by a belief that the conditions of the past were better than those of the present, and that steps should be taken to restore the remembered "golden age." The *inactive* attitude is characterized by a sound satisfaction with the present and a vested interest in maintaining the status quo; hence, any change is resisted. Because the reactive and inactive attitudes do not look toward *new* futures, neither has any need for scenario-based studies.

The *proactive* (also called *interactive*) attitude is distinguished by a sense of little satisfaction with the present and also an unwillingness either to recreate the past (which was also imperfect) or to accept an inevitable future. The pro-active attitude captures the belief that the future is the result of actions taken in the present. Implicit in this attitude is the need for positive visions of what society might achieve. The fourth attitude is the *preactive*; this is based on the presumption that the future cannot significantly be shaped by individuals or or-ganizations but instead will emerge from forces that are not readily controlled. Embedded in the preactive attitude is a desire for a forecast of what may happen in the hopes of exploiting new opportunities and avoiding pitfalls.

The split between proactive and preactive attitudes is reflected in what is, arguably, a broad distinction among kinds of scenario-based studies: *normative* studies, which seek to identify preferable futures; and *descriptive* studies, which aim to identify possible futures without regard for preference. Somewhat con-fusingly, many studies that use the expression "alternative futures" in the title involve normative futures,[18] whereas those that use "scenarios" usually consider descriptive futures.[19] These uses of the terms are distinct from Kahn's, who, as previously discussed, distinguished the ends and means of the future. For the sake of clarity, Kahn's suggested usage will be used throughout this book.

The divide between normative and descriptive approaches to future studies has been in evidence since scenario techniques were first developed in the years following World War II. Nearly 120 new nation-states came into existence dur-ing the 1950s and 1960s, and each had to reconcile its own traditional ways of life with newer international (but primarily Western) ideas of progress. Coun-tries that had been long established needed to look ahead and move beyond the wreckage tallied during two global conflicts and away from the past political ideologies, which had wrought havoc.[20] Moving forward was complicated by the lack of precedents by which to understand Cold War diplomacy. Further contributing to the uncertainty of life was the speed of scientific advances and the associated technological developments. The Allied victory revealed that, for the first time in history, the context of a conflict could radically change over a short span of time—even within the duration of a single war.[21] Some, such as the French philosopher and bureaucrat Gaston Berger, saw opportunities for inventing a better world.[22] Others, such as Kahn, saw uncertainties and, although they accepted that their tomorrows would be different from their yesterday, they doubted that the future could be successfully engineered.

Although the clear dichotomy between normative and descriptive scenarios is useful in a theoretical sense, and serves to orient a study, the distinction be-tween the two types is often blurred in practical application. In other words, some studies actively and freely mix normative and descriptive elements.[23] Even those authors who strive for methodological consistency also stray from purity for pragmatic reasons. Normative scenarios are overtly goal driven and value laden, but often include assumptions that are perceived to be value neutral relative to the study at hand. Conversely, descriptive scenarios are intended to be value neutral but, at the same time, are guided by the motivations of their authors. Although not intentional, implicit preferences may be conflated with beliefs of causality and distort the perception of the viability of a given future.[24] (For example, Michael Marien has argued that Kahn's work is biased by an implicit conservative ideology.[25] Of course, one could counterargue that the bias is in this interpretation of Kahn's work, not in its creation.)

Complete freedom from value-based assumptions is a difficult, if not impossible, task, and a high degree of self-consciousness is perhaps the best approximation to objectivity that the scenario creator and user may have. In addition, the distinction between normative and descriptive approaches can break down after the fact, as completed descriptive scenarios can be later judged as to their desirability.[26]

ATTITUDES AND STRATEGY

The motivation for exploring the future provides a first step in understanding the preconditions of a scenario-based study, but the sometimes vague boundary between the proactive attitude toward change (with associated normative scenarios) and the proactive attitude toward change (with associated descriptive scenarios) requires further refinement. Furthermore, irrespective of what a decision maker wants to occur, can he or she effect the desired results? More specifically, what is the range of available actions and what is the reach across time and space? The need to qualify proactive and proactive attitudes within a study and to account for varying degrees of agency leads to a consideration of the use of scenarios and, more broadly, an orientation to change.

Both normative and descriptive scenarios are used as an aid to decision making, but they are used in different ways. In normative scenario studies, the scenarios are themselves broad plans for the future, and the decision is that of which future to implement. In descriptive scenario studies, the scenarios are alternative conditions in which a more localized decision is compared; the better the decision does across the set of scenarios, the more robust the option is to future uncertainties.

The interrelated issues of agency and scenario use can be better understood through different notions of *strategy*—a word that is widely employed but not always with the same meaning. Roger Evered provides a useful comparison of the term as it is used in well-known texts from business, the military, and futures research.[27] The understanding of strategy in each of these three fields is, of course, both more varied and more complex than can be captured in a single work. But, divorced from their respective professional contexts, these distillations can serve as three generalized models of strategic thought. In turn, each of these strategic models incorporates a different attitude toward change and calls for different uses of scenarios.

In Kenneth Andrews' *The Concept of Corporate Strategy,* strategy is the domain of executives whose task is to identify opportunities, chart viable options, and achieve profits in the context of an open market and alongside the actions of rival peers.[28] Plans outline actions that can be implemented within the company by upper management. Significantly, the overall market in which the plan will be executed is not affected by the plan, although the company may capture a greater percentage of sales or increased profit margins on existing sales. A plan may also act to tap a currently available but not previously recognized segment of the market. Given the lack of influence on factors beyond the internal operations of the firm itself, the business strategy model reflects a preactive attitude toward change, and the scenarios developed for corporate clients are generally descriptive in character. For the chief executive officer, scenarios can model alternative future market contexts in which proposed plans can be tested. Indeed, such scenarios are sometimes described as "wind tunnels" to understand the consequences of potential internal actions.[29]

For the military, Basil H. Liddell-Hart's *Strategy* frames the issue as the *direction* of power so as to create advantageous situations in which a decision is achieved with the minimum exertion of force.[30] Strategic operations take place before the battle and are distinguished from tactical maneuvers which are the *application* of power during the fray. In this model, strategy is played out through attempts to change the underlying preconditions of a battle. The general can not only organize the forces and materiel under command (akin to the business executive managing employees and equipment); he or she can also reshape parts of the terrain that will be occupied. To be sure, some elements of the would-be battlefield are beyond control—such as the weather and the number of opponents—but hills can be fortified, streams can be bridged, and fields can be mined. In assuming that some aspects of the situation which are beyond the immediate confines of the organization can be altered toward achieving preferred outcomes, the military strategy model reflects a partially proactive and partially preactive attitude toward the future, and associated scenarios would hence be partly normative and partly descriptive.

Finally, for the futurist, in *On Learning to Plan* Don Michael describes strategy as a shared responsibility of all stakeholders and consisting of collectively appreciating change and making collaborative decisions to achieve a better standard of living for all.[31] In this model of strategy, plans are efforts to recreate the world in total and all aspects of society are considered to be subject to change and improvement. In contrast to the business model, which tests plans of action against scenarios, for the futurist scenarios *are* the plans of action. The attitude toward change is proactive and the associated scenarios are normative in character.

SCENARIOS AND THEIR USES

Literally, scenarios are "story lines" that allow people to understand the "flow" of events and from which we can examine and question the constants, trends, and shifts that are taking place. It seems useful to recall that the roots of both the words *history* and *story* spring from the same Greek word *historia*. Some languages, such as French, retain a single word for what others, such as speakers of English, differentiate with two. But of course, in policy debate, a scenario is not just *any* story; it is one that has been created to help examine and question choices for the future.

At their worst, scenarios help provide predetermined solutions to preconceived outcomes. Many who are engaged in intelligence work, gaming analysis, or "futurist" projections can easily fall victim to narrowly defined limits and acceptable outcomes. At their best, however, scenarios help distinguish among *threats, vulnerabilities,* and *risk*; scenarios may well be the best possible mechanisms to address the complex relationship among seemingly unrelated factors. The challenge for strategic planners is to help decision makers understand what the future security environment might look like, to affect their perceptions—in essence, to help them "reperceive." Pierre Wack, who gained some fame as a strategic planner during the oil crises of the 1970s with his ability to get the senior executives in Shell Oil to understand what might happen in the energy business, wrote the following in the *Harvard Business Review* some years later:

Scenarios deal with two worlds: the world of facts and the world of perceptions. They explore the facts but they aim at perceptions inside the heads of decision makers. Their purpose is to gather and transform information of strategic significance into fresh perceptions. This transformation process is not trivial—more often than not it does not happen. When it works, it is a creative experience that generates a heartfelt "Aha!" from you . . . [decision makers] and leads to strategic insights beyond the mind's previous reach.[32]

In short, to think and act effectively in an uncertain world, people must learn to *reperceive*—to question their assumptions and their understanding about the way the world works. By questioning those assumptions and rethinking the correct way to operate under uncertainty, we often see the world more clearly than we otherwise would. Wack summarized his goals as a strategic planner and developer of scenarios by stating:

I have found that getting to that [decision makers'] "Aha!" is the real challenge of scenario analysis. It does not simply leap at you when you've been presented all the possible alternatives. . . . It happens when your message reaches the microcosms of decision makers, *obliges them to question their assumptions about how their . . . world works, and leads them to change and reorganize their inner models of reality* [emphasis added].[33]

Former secretary of state Colin Powell, when he was Chairman of the Joint Chiefs of staff during the first Bush and the Clinton administrations, often valued such analyses as setting the context for a "strategic conversation" so that real, and often difficult, decisions could be made about the future. Scenarios help decision makers select alternative courses of action.

ALTERNATIVE SCENARIO APPROACHES: CLIMATE CHANGE AND HUMAN IMPACT

Scenarios, although acknowledged as useful tools for considering alternatives, are perhaps not well understood in how alternative methodological approaches used in creating them can also prove significant. One possible reason for this misperception—or conflicting understanding—is that those who practice scenarios often don't agree on the factors most critical to presenting coherent, plausible narrative scenarios.

Pierre Wack and Peter Schwartz, both former colleagues at Royal Dutch/ Shell in the 1970s, illustrated some of the more well-known alternate approaches to scenarios. Schwartz, the more prescriptive of the two, insists on identifying and dissecting three discriminating factors that lie at the heart of "understanding" the scenario process; he names these exploration factors *driving forces, predetermined elements,* and *critical uncertainties.* Wack, by contrast, advocates a looser structure to scenario construction and refuses to give overly precise definitions to discrete aspects, or elements, of the story line. This refusal to identify or separate specific aspects of the story suggests that it could be dangerous, even trivial, to reduce a coherent, plausible narrative to bare bones. Instead of looking only at the skeleton, we must also examine the flesh and blood of the story line in its integral wholeness. As such, the scenario reader (and scenario writer) open themselves to the complex interdependence among elements of a story and de-emphasize focusing on specific definitions.

At their best, scenarios provide *alternative projections* and possibilities for the future. Wack notes: "Scenarios serve two purposes. The first is protective—anticipating and understanding risk. The second is entrepreneurial—discovering strategic options of which one was previously unaware."[34] Creating and understanding scenarios aids better recognition of plausible—and sometimes implausible—outcomes. Ideally, they allow one to act on and better plan for potential outcomes in advance.

In the 1980s, for example, few in the profession of assessing the long-term global security environment forecast the demise of the Soviet Union. (Those who did were almost always ridiculed within their organizations.) Most conducting assessments and research saw the Cold War trends of the previous four decades as continuing indefinitely. Beginning with the fall of the Berlin Wall in 1989, and later with the Soviet Union's collapse, the U.S. defense establishment found itself in a significant force drawdown and witnessed the cancellation of countless billions of dollars of planned purchases.

Equally, though many strategic assessments at the beginning of the twenty-first century focused on American vulnerabilities and the potential danger of "asymmetric" warfare, these assessments seriously underestimated the damage that dedicated terrorists could inflict on the United States and the world. Scenarios that were designed, in other strategic assessments, were self-fulfilling—or, more appropriately, self-justifying—particularly in terms of defense budget and resource allocations. But then September 11, 2001, occurred.

Often, and probably naturally, decision makers prefer the illusion of certainty to understanding risk and realities. But the scenario "builder" and analyst should, at best, strive to shatter the decision maker's confidence in his or her ability to look ahead with certainty at the future. Scenarios should allow a decision maker to say, "I am prepared for whatever happens," because we have thought through complex choices with a knowledgeable sense of risk and reward.[35]

All too often, scenarios—just as larger framings of security issues themselves—center almost exclusively on threat-based analysis. In defense planning scenarios, analysis is driven by assessments of current or postulated threats or enemy capabilities and determines only the amount and types of force needed to defeat an adversary. (Thus, defense "transformation" attempts within the Pentagon to develop capabilities-based planning often seek to avoid the pitfalls and limits of threat-derived scenarios.)[36]

Scenarios that address climate change and human impact, nonetheless, must look well beyond current evaluations of threats. Accomplishing this is a difficult but essential challenge, if decision makers are to come to any informed, perceptive conclusions for the future. In short, as regards our specific interest in climate change and its human impact, a focus away from direct threats and more toward creeping, long-term vulnerabilities seems necessary.

As regards climate change, how much do we know is natural (even cyclic) occurrence, and how much is directly related to anthropogenic causes such as pollution, depletion of resources, and environmental footprints? The answer remains unclear, yet the critical recognition, over recent history, is that events are accelerating rapidly, and generally the impact on human outcomes could prove lethal if nothing is done. Thus, scenarios provide an opportunity to reckon with alternative conditions. Within the previously discussed Schwartz and Randall assessment, for example, three conditions that previously existed in history were offered as possible outcomes for rapid climate change and its resulting impact on security conditions:

- **Extreme:** Return to the Younger Dryas, ending 9600 B.C. The Younger Dryas was the most significant rapid climate change event that occurred during the last deglaciation of the North Atlantic region. Although difficult to picture perhaps today, during this period icebergs were present south of Portugal, Greenland was 15 degrees C colder than today, increased dust from Asian deserts was present in the atmosphere, and there was increased glaciation in mountainous altitudes throughout the globe. Prevailing theory suggests that a major reduction, and perhaps even a shutdown, of thermohaline circulation occurred due to fresh water flowing into the North Atlantic from deglaciation in North America.[37]
- **Feasible:** A return to the "Little Ice Age," which was characterized by hard winters, violent storms, and drought, such as occurred between 1300 and 1850 throughout Europe and the North Atlantic. Colder weather severely affected agriculture, health, economics, social strife, emigration, and even art and literature. Increased glaciation and storms devastated those who lived near glaciers and the sea. Furthermore, as Mandia notes, the growing season changed by 15 to 20 percent between the warmest and coldest times of the millennium—directly affecting food production, especially crops highly adapted to use the full-season warm climatic periods. Climate changes directly affected human security outcomes.[38] Figures 4.2 and 4.3 show the

Figure 4.2
Prices of wheat in Dutch guilders per 100 kilograms in various countries versus time. Source: *H. H. Lamb, Climate, History and the Modern World* (London: Methuen, 1995).

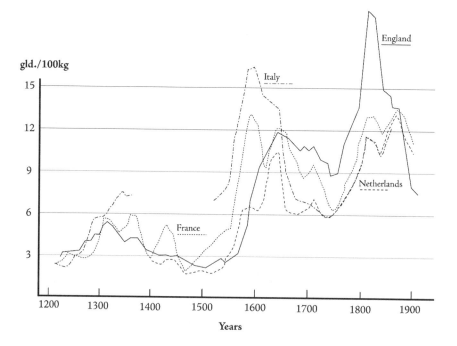

Figure 4.3
Price of rye in Germany versus time expressed as an index. Source: H. H. Lamb, *Climate, History and the Modern World* **(London: Methuen, 1995).**

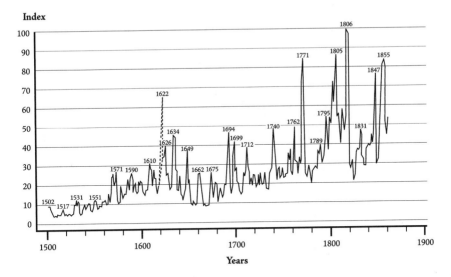

price of wheat and rye, respectively, in European countries during the Little Ice Age. Figure 4.4 offers a chronology of dearth and famine in Scotland during the Little Ice Age.

- **Reasonable:** Cold, dry, windy effects across much of the Northern hemisphere—caused by "conveyor collapse" that occurred 8,200 years ago. Since a good bit of data exists regarding this event, much attention has equally focused on the possibility of the slowing of thermohaline circulation. Such circulation, nonetheless, is critical to environmental conditions; the United Kingdom, for example, though roughly similar in latitude to Labrador, possesses a markedly different environment, thanks to such circulation. The conveyor belt principle is relatively simple: As sea ice freezes, salt drains from the pores of the ice. Yet, owing to both evaporation and heat loss, water from the tropics becomes denser as it drifts toward the Arctic and Greenland. Salty water, because it is more dense, sinks to the ocean floor. As a result, more warm water is drawn from the tropics to the poles, moving heat and affecting human environmental conditions around the globe.[39]

Although the scenarios on their own may indeed seem extreme as possible conditions leading to rapid climate change in our time, their value lies in the considerable debate their potential for human impact has generated. Lamentably, much of that debate took place in the media; little practicable or practical policy action has taken place since the Pentagon released this scenario report. Yet, today, scientists daily discover emerging trends about the possible "acceleration" in the loss of the Greenland ice sheet, outcomes that constantly force us

Figure 4.4
Dearth and famine in Scotland during the Little Ice Age. Source: H. H. Lamb, *Climate, History and the Modern World* (London: Methuen, 1995).

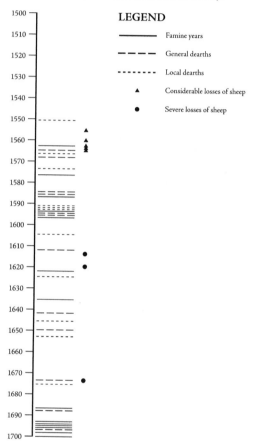

back to thinking the unthinkable about abrupt climate change, about asking the question "What if . . . ?"

NOTES

1. Lewis Carroll, "Pig and Pepper," *Alice's Adventure in Wonderland,* Millennium Fulcrum 3.0, at www-2.cs.cmu.edu/People/rgs/alice-ftitle.html.

2. Peter Schwartz and Doug Randall, *An Abrupt Climate Change Scenario and Its Implication for United States National Security* (Written for the Office of Net Assessment, within the Office of the Secretary of Defense: Washington, DC, October 2003). http://www.ems.org/climate/pentagon_climatechange.pdf.

3. A. Kleiner, *The Age of Heretics: Heroes, Outlaws, and the Forerunners of Corporate Change* (New York: Currency Doubleday, 1996).

4. H. Kahn and A. J. Wiener, *The Year 2000: A Framework for Speculation on the Next Thirty Years* (New York: Macmillan, 1967), p. 6.

5. See, for example, E. Cornish, *Futuring: The Exploration of the Future* (Washington, DC: The World Future Society, 2004); N. C. Georgantzas and W. Acar, *Scenario-Driven Planning: Learning to Manage Uncertainty* (Westport, CT: Quorum Books, 1995); R. J. Lempert, S. W. Popper, and S. C. Bankes, *Shaping the Next One Hundred Years: New Methods for Quantitative, Long-Term Policy Analysis*, RAND report MR-1626 (Santa Monica, CA: RAND Corporation, 2003); G. Ringland, *Scenario Planning: Managing for the Future* (New York: John Wiley, 1998); Peter Schwartz, *The Art of the Long View: Paths to Strategic Insight for Yourself and Your Company* (New York: Currency Doubleday, 1996, first published 1991); P. Schwartz and J. A. Ogilvy, "Plotting your scenarios," in *Learning from the Future: Competitive Foresight Scenarios,* ed. L. Fahey and R. M. Randall (New York: John Wiley, 1991), pp. 57–80; K. van der Heijdn, *Scenarios: The Art of Strategic Conversation* (New York: John Wiley, 1996); U. von Reibnitz, *Scenario Techniques,* trans. P. A. W. Rosenthal (New York: McGraw-Hill, 1987).

6. W. Bell, *Foundations of Future Studies: Human Science for a New Era, Volume 1— History, Purposes, and Knowledge* (New Brunswick, NJ: Transaction Books, 1997).

7. Kahn and Wiener.

8. S. Chatman, *Story and Discourse: Narrative Structure in Fiction and Film* (Ithaca, NY: Cornell University Press, 1980).

9. See van der Heijdn, 1996.

10. Ibid., Schwartz and Ogilvy, 1998.

11. Thomas S. Kuhn, *The Structure of Scientific Revolutions,* 3rd ed. (Chicago: University of Chicago Press, 1996). The use of the term *cliché* by no means casts a pejorative slant on Kuhn's original work, which was innovative and startling in its originality; rather, the use and abuse of "paradigm shift" has in many ways been cast in such a wide net as to have separated from earlier intended meanings.

12. One of several media accounts of the "strangelet scenario" can be found at: http://www.phy.bnl.gov/users/inthenews/nd121200.html.

13. Martin Rees, *Our Final Hour: A Scientist's Warning: How Terror, Error, and Environmental Disaster Threaten Humankind's Future in This Century—On Earth and Beyond* (New York: Basic Books, 2003), pp. 120–21, quoted in Richard A. Posner, *Catastrophe: Risk and Response* (New York: Oxford University Press, 2004), p. 30.

14. John Marburger, "Synopsis of Committee Report, October 6, 1999." http://www.phys.utk.edu/rhip/Articles/RHICNews/BNL_rhicreport.html. Quoted in Posner, p. 31.

15. Posner, pp. 31–32.

16. Ibid., pp. 33.

17. Russell L. Ackoff, *Creating the Corporate Future: Plan or Be Planned For* (New York: John Wiley, 1981).

18. For example, see L. R. Beres and H. R. Targ, *Constructing Alternative World Futures: Reordering the Planet* (Cambridge, MA: Schenkman, 1977).

19. For example, see G. Ringland, *Scenario Planning: Managing for the Future* (New York: John Wiley, 1998).

20. W. Bell, 1997.

21. H. Kahn and I. Mann, *Techniques of Systems Analysis*, RAND report RM-1829 (Santa Monica, CA: RAND Corporation, 1956).

22. A. Cournand and M. Levy, eds, *Shaping the Future: Gaston Berger and the Concept of Prospective* (New York: Gordon and Breach, 1973).

23. E. Jantsch, *Technological Forecasting in Perspective* (Paris: OECD, 1967).

24. A. M. Washburn and T. E. Jones, "Anchoring Futures in Preferences," in *Forecasting in International Relations: Theory, Methods, Problems, Prospects,* ed. N. Choucri and T. W. Robinson (San Francisco: W. H. Freeman, 1978), 95–115.

25. Michael Harien, "Herman Kahn's Things to Come," *The Futurist* 7:1 (1973): 7–15.

26. W. Bell, 1997.

27. Roger Evered, "So What *Is* Strategy?" *Long Range Planning* 16:3 (1983): pp. 57–72.

28. Kenneth R. Andrews, *The Concept of Corporate Strategy* (Homewood, IL: Dow-Jones Irwin, 1970).

29. van der Heijden, 1996.

30. Basil H. Liddell-Hart, *Strategy* (New York: Praeger, 1965).

31. Don Michael, *On Learning to Plan—and Planning to Learn* (San Francisco: Jossey-Bass, 1973).

32. Pierre Wack, "Scenarios: Shooting the Rapids: How Medium-Term Analysis Illuminated the Power of Scenarios for Shell Management," *Harvard Business Review,* November–December 1985, p. 140. Other general works regarding scenarios and strategic implications include Robert Jervis, *System Effects: Complexity in Political and Social Life* (Princeton, NJ: Princeton University Press, 1997); Kees van der Heijdn, *Scenarios: The Art of Strategic Conversation* (Chichester, UK: John Wiley & Sons, 1996); Seymour J. Deitchman, *On Being a Superpower: Scenarios for Security in the New Century* (Boulder, CO: Westview Press, 2000); and Gill Ringland, *Scenario Planning: Managing for the Future* (New York: John Wiley & Sons, 1998).

33. Ibid. Portions of this section have been adapted from P. H. Liotta and Timothy E. Somes, "The Art of Reperceiving: Scenarios and the Future," *Naval War College Review* 56, no. 4 (Autumn 2003), http://www.nwc.navy.mil/press/Review/2003/Autumn/cy1-a03.htm.

34. Wack, p. 146.

35. Peter Schwartz, "The Art of the Long View," in *Strategy and Force Planning,* ed. Strategy and Force Planning Faculty, 2nd ed. (Newport, RI: Naval War College, 1997), chap. 3. Reprinted by permission from Peter Schwartz, *The Art of the Long View* (New York: Doubleday, 1996), pp. 3–10 and 105–23.

36. See John F. Troxell, *Force Planning in an Era of Uncertainty: Two MRCs as a Force Sizing Framework* (Carlisle Barracks, PA: Strategic Studies Institute, 1997) for a detailed discussion of both threat-based scenario development and capabilities-based planning. See also Henry Bartlett, G. Paul Holmes, and Timothy E. Somes, "The Art of Strategy and Force Planning," in *Strategy and Force Planning,* ed. Strategy and Force Planning Faculty, 3rd ed. (Newport, RI: Naval War College, 2000), chaps. 2 and 26–30, for alternative approaches to force planning.

37. http://www.agu.org/revgeophys/mayews01/node6.html; http://en.wikipedia.org/wiki/Younger_Dryas.

38. Drawn from information compiled by Scott A. Mandia, Suffolk County Community College, on "The Little Ice Age in Europe." Data referenced is drawn from: H. H. Lamb, *Climate, History and the Modern World* (London: Methuen, 1995); http://www2.sunysuffolk.edu/mandias/lia/little_ice_age.html.

39. Elizabeth Kolbert, "Annals of Science: The Climate of Man—I," *New Yorker,* April 25, 2005, p. 69.

— 5 —

"Let's Scare Jessica to Death": Mind-Sets, the Fear Factor, and Climate Change

Man's role is uncertain, undefined, and perhaps unnecessary.

Margaret Mead

In this chapter, we mean to address more than a consideration of how alternative scenarios are built. Rather, we are equally interested in the *mind-sets*—the mental maps, as it were—of decision makers. Although it may be a cliché, it is also an evident certainty that how we view the world subtly but vitally determines how we act in it. All of us must have "mental maps"; all of us must recognize the constraints, however, that such mapping imposes on our own recognitions.

One of the more intriguing approaches to at least conceptually addressing strategic scenarios under uncertainty appeared in the *Harvard Business Review* in the late 1990s. Titled "Strategy under Uncertainty," the essay addressed how executive decision makers might best address new forms of strategic change, while equally admitting that traditional strategic planning processes were unlikely to bear much fruit during times of significant shift.[1] Moreover, the essay suggests, decision makers tend to view uncertainty with binary polarity: either predictions about the future are certain, or they are completely uncertain.[2] Risk-averse decision makers tend to do little in uncertain environments, choosing instead to defer to decision paralysis, to focus on reengineering or program streamlining or cost-reduction efforts.[3]

LEVEL 1: THE RUSE OF UNCERTAINTY

While acknowledging that residual uncertainty will *always* exist, the authors chose instead to consider four levels of uncertainty. In Level 1, shown in figure 5.1, any form of residual uncertainty is not relevant to decisions necessary to be made. Just as, for example, in the Cold War, the enemy and military force composition were (supposedly) well known, the necessary adjustments and targeting mechanisms, along with the required military force structure, were *all* straight line projections. Graphically, this project would look like:

Essentially, a Level 1 future is a linear process. The future itself is clear enough; one need only apply the necessary resources to achieve a desired direction—or deflect an undesirable outcome. Level 1, in other words, is *not* uncertain at all.

In our discussion in chapter 4 regarding scenarios and alternative futures, we discuss the "surprise-free" alternative future (which was marked "sf" on figure 4.1). Figure 5.1 is a specific representation of this "business as usual" expectation; the future can be anticipated—and for some in a sense the future can be *controlled*—if

Figure 5.1
Level 1 Uncertainty: A Clear Enough Future
A Single Forecast Precise Enough for Determining Strategy. Reprinted by permission of *Harvard Business Review*. Adapted from "Strategy Under Uncertainty" by Hugh Courtney, Jane Kirkland, and Patrick Viguerie, 68:3 (November–December 1997), 79–91. Copyright (c) 1997 by the Harvard Business School Publishing Corporation; all rights reserved.

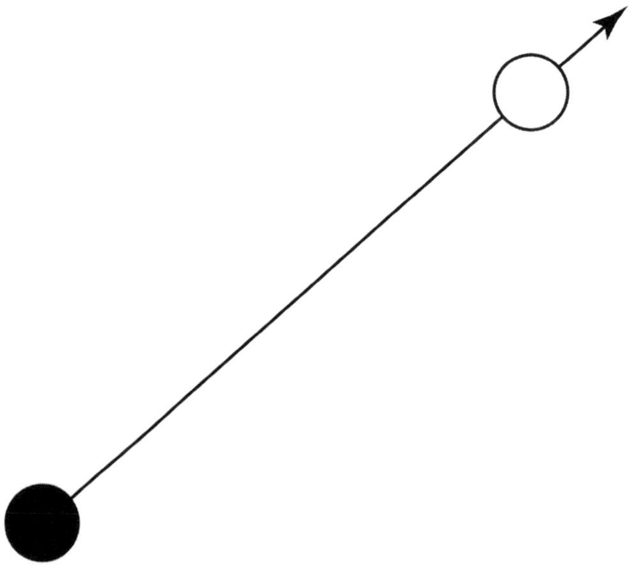

there are no significant changes in the social, political, economic, technical, or environmental aspects of the world. In part, such stability precludes changes *in kind*.

As regards climate change, and a single-mindedness regarding its certainty, a prototypical example appeared in the Winter 2003 issue of the *Naval War College Review:*

The fraction of carbon dioxide (CO_2) in the atmosphere has been slowly but steadily increasing since systematic observations began a century ago. Little concern was evident until the mid-1980s, when some researchers suggested that CO_2 would warm the atmosphere by absorbing infrared radiation emitted by the earth. Environmentalists soon joined on an international scale to clamor for stringent controls on the sources of CO_2. The result was the Kyoto Protocol. ...

The protocol, which is both lengthy and complex, requires large reductions in CO_2 emissions. (The United States would have to reduce CO_2 emissions to a level 7 percent below that of 1990 by the years 2008 to 2012—this despite the steady growth of the U.S. population and the phase-out of nuclear power generation.) The "Third World," including the giants China and India, is exempt. Despite this exemption, Third World countries would, under the terms of the Protocol, accrue "credits" for emissions, which they could sell to the "First World." In other words, the protocol would become an instrument for transfer of wealth from nations such as the United States to Third World elites, a sort of international welfare scheme under a misleading name.

Carbon dioxide molecules can warm the atmosphere through changes ("excitation") in their vibrational and rotational properties. (For CO_2, such excitation occurs in the infrared part of the electromagnetic spectrum, in which the earth is an efficient emitter. Heating of the atmosphere occurs by transfer of energy from CO_2 to air molecules via molecular collisions.) This compound is not the only atmospheric greenhouse gas, for several others, such as nitrous oxide (N_2O) and methane (CH_4), are also covered by the Kyoto Protocol. However, water vapor, which cycles through the atmosphere in about a week via evaporation from oceans, lakes, and rivers followed by condensation and precipitation, is far and away the most important greenhouse gas, because it is plentiful in the atmosphere and it strongly absorbs infrared radiation emitted by the earth. Absence from the atmosphere of water vapor would make the entire earth like the Sahara Desert—or, to state it more dramatically, like Mars. In contrast, CO_2, with a cycle duration of, according to recent analysis, thirty to fifty years, is much less plentiful and absorbs infrared radiation more weakly than does water vapor. The major removal mechanisms for CO_2 are absorption by vegetation and the oceans....

We know from geologic records, tree rings, and human records that the mean temperature of the atmosphere has varied markedly during the past million years. The most glaring aspect of the record is the series of glacial periods, at least partially associated with the earth's orbital characteristics, which last on average about ninety thousand years, with "interglacial" periods of about eleven thousand years. A more recent feature of the record is the "little Ice Age," which lasted from the end of the fourteenth century until about 1850, when began a gradual temperature rise that essentially ended in 1940. This period, which was characterized by low agricultural productivity and frequent famines, may have been due in part to reduced energy output by the sun; sunspot activity was abnormally low during much of the period of low temperature.

The rise in mean atmospheric temperature during this century is often cited as evidence of the warming effect of CO_2. Those who cite this "evidence" fail, however, to mention that nearly all of the warming occurred before 1940, as the earth recovered from the "little Ice Age."

What of the most recent record? Atmospheric temperature measurements are routinely made at airports, in urban areas, at sea … They are also made by balloon-borne radiosondes and (since 1978) by satellites. While the surface measurements do show a small temperature rise (about 0.6° F since 1980), they are contaminated by the so-called "urban heat island" effect. Urban areas and airports have been emitting greater and greater amounts of heat energy as a result of growing human activity that has nothing to do with the greenhouse effect. Corrections applied to the surface data are unreliable, because a large degree of estimation is involved.

Balloon-borne radiosonde and satellite measurements of the temperature of the "free atmosphere" (i.e., at heights that would capture any heating caused by CO_2) are far more reliable. Although the records are characterized by a wild oscillation, one can compute a trend line using standard spreadsheet methods. The observed temperature change decreases slightly with time for the balloon data and is essentially zero for the satellite measurements. The National Academy of Sciences has recognized the conflict between satellite and surface temperature measurements as a major problem with no known explanation.

U.S. energy consumption in 2000 was more than 10 percent larger than in 1990. Thus a 7 percent decrease from 1990 consumption would really mean either a 15 to 20 percent drop below 2000 levels or a very substantial increase in tax rates to purchase "credits" from Third World countries. Despite claims by treaty proponents to the contrary, such a reduction in use would have severe economic consequences, because of the strong dependence of our economy on energy. William Nordhaus of Yale University has calculated the cost to the world economy of "stabilizing" the climate to be $12.5 trillion (1989 dollars). Since the United States consumes about 25 percent of the world's fossil fuels, the cost to this nation would be in excess of three trillion dollars, an enormous stress to place on its economy. The national defense would also suffer, because of the enormous fuel requirements for training the armed forces, not to mention those for combat, as in Afghanistan. All this when there is *no credible evidence* for global warming due to carbon dioxide emissions.[4]

On the surface, at least, this does appear to be quite an impressive argument. The one (perhaps myopic) claim that we are on a straight-line projection, and should not divert resources for strategic benefits, because of faulty scientific measurement of CO_2 emissions, fails to recognize a fairly significant alternative proof, nonetheless. Ice cores from Antarctica effectively trace back more than four glacial cycles. Since gas samples themselves become trapped within the core as tiny bubbles, scientists have been able to demonstrate that CO_2 levels are far higher than at any time during the last 420,000 years. Furthermore, roughly half the temperature variance between cold and warm periods can be linked to changes in the concentration of greenhouse gases.[5] Arguably, the mind-set of Level 1 regarding climate change and human impact is anything *but* a certain, straight-line project.

Moreover, this analysis refers to balloon-borne radiosonde and satellite measurements as more accurately measuring the temperatures of the free atmosphere.[6] Recent work on these measurements, nonetheless, has shown that data used in these measurements—which for years have buoyed the arguments of climate change skeptics—were actually based on faulty analysis. In 2005, separate studies published in *Science* show that this atmosphere is indeed warming, whereas satellite and weather balloon data had supposedly shown the opposite. According to Steven Sherwood, a geologist at Yale and lead author of one of the studies, his examination of radiosondes (which are capable of making direct measurements of the atmosphere), showed differences in the nature of measurements taken twice in a 24-hour period, once during the day and once at night.[7] Sherwood noted that older radiosonde instruments were not as shielded from direct sunlight as more recent models and thus (because they were affected by direct sunlight) showed warmer temperatures during the day. Older measurements thus collected temperature readings higher than actually true, with an overall effect that appeared to suggest that the troposphere was cooling. When analyzed together, older and newer radiosonde data showed significant discrepancies in daytime measurements.[8]

Similarly, until recently, the research team under the direction of Roy Spencer from the University of Alabama was the only group to previously (in 1992) analyze satellite data on the troposphere. In another study published in *Science,* Carl Mears and Frank Wentz, from the California-based Remote Sensing Systems, reviewing the same data and discovered an error in Spencer's analysis technique; after correcting for the error, newer analysis again shows that the troposphere is warming, (not cooling, as had been previously claimed).[9] In the third study regarding weather-balloon—borne radiosonde and satellite data, Ben Santer, an atmosphere scientist at the Livermore National Laboratory in California, applied the newer Mears and Wentz analysis to illustrate how the "new" data was consistent with climate models and theories.[10]

Finally the extensive excerpt (which is meant both as an illustration of Level 1 uncertainty and a demonstration of how Level 1 conditions of certainty are far less frequent than might at first blush seem) suggests that "U.S. energy consumption in 2000 was more than 10 percent larger than in 1990."[11] When President Clinton left office, greenhouse CO_2 emissions were 15 percent higher than they had been in 1990.[12] As of this writing, U.S. emissions are nearly 20 percent higher than they were in 1990.[13] The *result* of those emissions is indeed related to energy consumption, nonetheless. The average American, for example, produces the same greenhouse-gas emissions as 4.5 Mexicans, 18 Indians, or 99 Bangladeshis. Although clearly the economic impact of modifying consumption and emission patterns would be significant, the cost of not adapting would result in a climate "boomerang" that would look nothing like a Level 1 "clear enough future."

LEVEL 2: LIMITING UNCERTAIN OUTCOMES

In Level 2 uncertainty, decision makers enter a pathway of "branches and sequels," which the authors term "Alternative Futures." Although no clear outcome is certain, scenario analysis may help determine both outcome probabilities and outcome damage (see figure 5.2).

Although not clearly addressed in the essay, Level 2 uncertainty may help decision makers acknowledge and better deal with risk—especially in light of our minimalist definition of risk as *the ability to expose oneself to damage during the process of change and the resilience to be able to sustain oneself during such change.* Although pathways and outcomes are more sophisticated than in Level 1, there are still some manageable "shaping" mechanisms that can be applied to deal with outcomes.

There are clear advantages, nonetheless, of simply connecting the logical standard "If A leads to B, what mitigating responses—whether C or iterative response of C, D, E..."—might best address Level 2 uncertainty?" A powerful example of this response is found in the monograph *The Death of Environmentalism: Global Warming Politics in a Post-Environmental World.* Although the tract itself is clearly meant as a polemic, the authors argue (with some justification) that environmentalists have *reverted* to Level 1 analysis by too narrowly focusing on what constitutes "environmental policy making":

... [I]n their public campaigns, not one of America's environmental leaders is articulating a vision of the future commensurate with the magnitude of the crisis. Instead they are promoting technical policy fixes like pollution controls and higher vehicle mileage standards—proposals that provide neither the popular inspiration nor the political alliances the community needs to deal with the problem.

By failing to question their most basic assumptions about the problem and the solution, environmental leaders are like generals fighting the last war—in particular the war they fought and won for basic environmental protections more than 30 years ago. It was then that the community's political strategy became defined around using science to define the problem as "environmental" and crafting technical policy proposals as solutions.

The greatest achievements to reduce global warming are today happening in Europe. Britain has agreed to cut carbon emissions by 60 percent over 50 years, Holland by 80 percent in 40 years, and Germany by 50 percent in 50 years. Russia [ratified] Kyoto. And even China—which is seen fearfully for the amount of dirty coal it intends to burn—recently established fuel economy standards for its cars and trucks that are much tougher than ours in the US.[14]

The authors, nonetheless, offer reasonable Level 2 outcome actions and policy responses, even as they suggest that such basic strategic planning has failed to take place:

What do we worry about when we worry about global warming? Is it the refugee crisis that will be caused when Caribbean nations are flooded? If so, shouldn't our focus be on building bigger sea walls and disaster preparedness?

Figure 5.2
Level 2 Uncertainty: Alternative Futures
A Few Discrete Outcomes that Define the Future.
Reprinted by permission of *Harvard Business Review.*
Adapted from "Strategy Under Uncertainty" by Hugh
Courtney, Jane Kirkland, and Patrick Viguerie, 68:3
(November–December 1997), 79–91. Copyright (c) 1997
by the Harvard Business School Publishing Corporation;
all rights reserved.

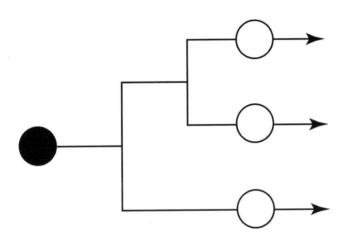

Is it the food shortages that will result from reduced agricultural production? If so, shouldn't our focus be on increasing food production?

Is it the potential collapse of the Gulf Stream, which could freeze upper North America and northern Europe and trigger, as a recent Pentagon scenario suggests, world war?

Most environmental leaders would scoff at such framings of the problem and retort, "Disaster preparedness is not an environmental problem." It is a hallmark of environmental rationality to believe that we environmentalists search for "root causes" not "symptoms." What, then, is the cause of global warming?

For most within the environmental community, the answer is easy: too much carbon in the atmosphere. Framed this way, the solution is logical: we need to pass legislation that reduces carbon emissions. But what are the obstacles to removing carbon from the atmosphere?

Consider what would happen if we identified the obstacles as:

• The radical right's control of all three branches of the US government.
• Trade policies that undermine environmental protections.
• Our failure to articulate an inspiring and positive vision.

- Overpopulation.
- The influence of money in American politics.
- Our inability to craft legislative proposals that shape the debate around core American values.
- Poverty.
- Old assumptions about what the problem is and what it isn't.

The point here is not just that global warming has many causes but also that the solutions we dream up depend on how we structure the problem.

The environmental movement's failure to craft inspiring and powerful proposals to deal with global warming is directly related to the movement's reductive logic about the supposedly root causes (e.g., "too much carbon in the atmosphere") of any given environmental problem. The problem is that once you identify something as the root cause, you have little reason to look for even deeper causes or connections with other root causes.[15]

Although their individual points may indeed be debatable, Shellenberger and Nordhaus do emphasize that dealing with the sequential and logical outcomes of events is essential in recognizing scenario potential and acting on valid, effective outcomes.

In another sense, however, resisting the opportunity to deeply consider potentially significant outcomes, counter to one's own position, could also be considered part of Level 2 uncertainty, in seeking to limit a few discrete outcomes that define the future. In that sense, the job of any bureaucrat (or diplomat) in seeking to control positions that run counter to the mind-sets of decision makers can be particularly difficult. Paula Dobriansky, U.S. Under Secretary of State for Global Affairs and Democracy during the G.W. Bush administration, consistently insisted that, regarding climate change, "Science tells us that we cannot say with any certainty what constitutes a dangerous level of warming and therefore what level must be avoided....We predicate our policies on sound science."[16]

When the science involves uncertainty, nonetheless, it becomes increasingly complex to define precisely the bases for "sound science"—unless one can speak with the absolute certainty of hindsight. While stressing that the United States does indeed take climate change seriously, Dobriansky explains that rather than engaging in the full protocol requirements of Kyoto (the complete official title is the Kyoto Protocol to the United Nations Framework Convention on Climate Change), the United States has engaged in 15 bilateral initiatives that address issues of global warming and climate change. When queried about when and if the administration would accede to mandatory caps on emissions, her common response reflected a Level 2 uncertainty mind-set: "We act, we learn, we act again."[17]

Admittedly, the addendum requirements to the Kyoto Framework Convention do impose stricter emission reduction requirements than for those in the so-called developing world; the European Union is expected to reduce greenhouse gas emissions 8 percent below 1990 levels by 2012, and the United States is set with a target of 7 percent below 1990 levels. The United States, of course, is one of only two developed nations to have rejected the Kyoto Protocol. (The other is

Australia.) It is therefore all the more significant that the United States (which, as noted, accounts for 34 percent of Annex I emissions), having rejected the protocol, was still unable to prevent it from coming into effect—since its approval had to come from countries responsible for at least 55 percent of those emissions.

Notably, Paula Dobriansky's sister Laura also worked on administration policies regarding climate change. Laura Dobriansky served as Deputy Assistant Secretary for National Energy Policy at the Department of Energy (where she managed the department's Office of Climate Change Policy) and previously worked on climate change issues as a lobbyist for a firm that represented Exxon Mobil. Although theories of inside manipulation and special interests might easily abound here, the world, again, is a bit more complex a place. Yet Under Secretary of State Dobriansky emphasized "the rhetoric of scientific uncertainty" as a consistent theme regarding climate change and human impact. Speaking at a November 2003 panel sponsored by the American Enterprise Institute, she remarked that "the extent to which the man-made portion of greenhouse gases is causing temperatures to rise is still unknown. Predicting what will happen fifty or one hundred years in the future is difficult."[18] That said, however, one could easily counterargue that—with the implications of global climate change, as various scenarios presented in this work have offered—there is every *reason* to deal in reasonable projection, even if accurate prediction is impossible.

In Senate testimony in November 2005, Under Secretary Dobriansky provided an address titled "U.S.-International Climate Change Approach: A Clean Technology Solution." In her testimony, she emphasized a joint global response to climate change, a focus on development agenda that linked economic growth, poverty reduction, and improved resource protection, while noting that technology is the "glue that can bind ... development objectives together." Finally, she noted that engaging private sector innovation and ingenuity was at least as important as government-to-government collaboration.[19] Two weeks after this testimony, Dobriansky headed the U.S. delegation to the Conference of the Parties (COP) to the United Nations Framework on Climate Change, held in Montreal from November 28 to December 9. At the conclusion of that session, despite the Bush administration's adamant resistance, the vast majority of developed states agreed to engage in a new set of binding limit requirements for greenhouse gas emissions that would take effect in 2012—in essence, setting a stage for a post-Kyoto agenda.[20]

It would seem that, once again without U.S. participation, the remainder of the world was ready to engage with greater levels of uncertainty, and to take actions to address that uncertainty.

LEVEL 3: THINGS BEGIN TO GET FUZZY

Level 3 uncertainty—and the shaping mechanisms necessary to respond to such uncertainty—require more nuance, more flexibility, and more adaptive postures. Rather than being able to "shape" the influencing agent (in our case,

Figure 5.3
Level 3 uncertainty: A Range of Futures
A Range of Possible Outcomes, but No Natural Scenarios. Reprinted by permission of *Harvard Business Review*. Adapted from "Strategy Under Uncertainty" by Hugh Courtney, Jane Kirkland, and Patrick Viguerie, 68:3 (November–December 1997), 79–91. Copyright (c) 1997 by the Harvard Business School Publishing Corporation; all rights reserved.

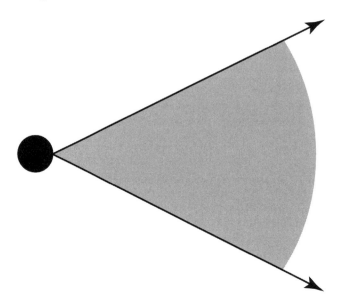

climate change) to lessen or mitigate outcomes (human impact), decision makers are able, at best, to only identify a range of possible outcomes (See figure 5.3).[21]

One example of an attempt to suggest the Level 3 range impact of climate change and its concomitant effect on a specific human impact factor (economic pressures) was published by the Congressional Budget Office in 2003. The report, titled *The Economics of Climate Change: A Primer*, effectively and repeatedly deals with how uncertainty involves a "range of futures." Two powerful examples of this range include those shown in figure 5.4 and figure 5.5, which address the uncertainty of future predictions regarding climate change outcomes.

The benefit of such Level 3 uncertainty and its usefulness in scenarios is that decision makers recognize the interconnectedness of often complex dynamics and the necessity to think in creative ways and to act through *inter*dependent means. The necessity for international protocols, the role of international institutions, and even the influence of private citizens and multinational businesses take on new importance in addressing a common challenge. Admittedly, the

Figure 5.4
Range of Uncertainty in Economic and Carbon Dioxide Emission Projections

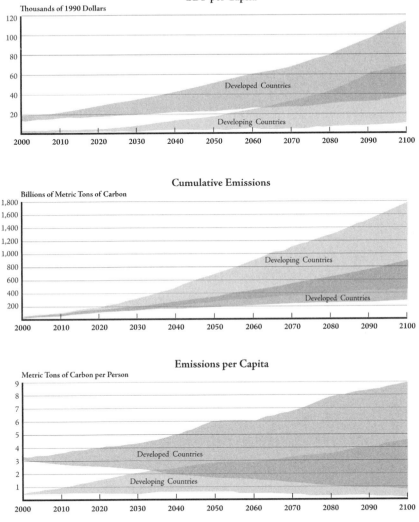

GDP per Capita

Cumulative Emissions

Emissions per Capita

Source: Congressional Budget Office, based on Nebojša Nakicenovićand Rob Swart eds., *Emission Scenarios* (Cambridge: Cambridge University Press, 2000).
Note: All emissions are from fossil fuels.

Figure 5.5
Historical and Projected Climate Change
(Average Global Temperature (C) Relative to 1986–1995 Average)

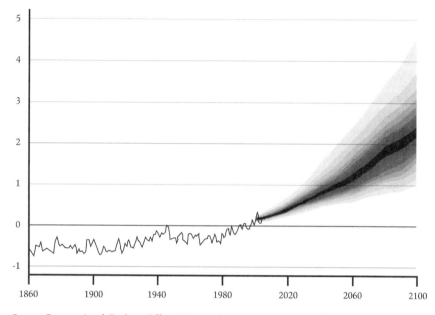

Source: Congressional Budget Office. Historical data are from the Hadley Centre for Climate Prediction and Research, available at www.met-office.gov.uk/research/hadleycentre/CR_data/ Annual/land+sst_ web.txt and described primarily in C. K. Folland and colleagues, "Global Temperature Change and Its Uncertainties since 1861," *Geophysical Research Letters* 28 (July 1, 2001): pp. 2621–24. The projection is based on data provided by Mort Webster, University of North Carolina at Chapel Hill, in a personal communication, December 11, 2002; the results are discussed in Mort Webster and colleagues, *Uncertainty Analysis of Climate Change and Policy Response*, Report no. 95 (Cambridge, MA: Massachusetts Institute of Technology Joint Program on the Science and Policy of Global Change, December 2002).

Note: The projection, which is interpolated from decadal averages beginning in 1995, shows the possible distribution of changes in average global temperature as a result of human influence, relative to the 1986–1995 average and given current understanding of the climate. Under the Webster study's assumptions, the probability is 10 percent that the actual global temperature will fall in the darkest area and 90 percent that it will fall within the whole shaded area. However, actual temperatures could be affected by factors that were not addressed in the study (such as volcanic activity and the variability of solar radiation) and whose effects are not included in the figure.

scientific "imprecision" that figure 5.5 provides—despite the best efforts to project possible ranges of temperature outcomes in the future without the benefit of completely accurate Level 1 prediction—leaves most, if not all decision makers in states of blurred confusion. In most instances, because decision makers must live in the "real world," they will not bother to act—or even respond. In short, policy demands choices to respond to; the wide array of temperature variance predictions in figure 5.5 leaves only allowance for consideration of a wide range of impact considerations, and little room for accurate knowledge of what choices to make, what means might be available to mitigate, or what instruments or methods exist to adapt to coming impact changes.

Michael Crichton, bestselling author of *Jurassic Park* and more recently of *State of Fear* (a novel in which ecoterrorists disrupt weather patterns and stir fears about global climate change), is a self-proclaimed climate change skeptic. He does not mince his words, either: "[T]he evidence for so many environmental issues is, from my point of view, shockingly flawed and unsubstantiated."[22] Indeed, in an address to the National Press Club on January 25, 2005, he was quite specific in his critique: "We are basing our decisions on speculation, not evidence. Proponents are pressing our views with more PR than scientific data."[23] He further specifically critiques language from the Third Assessment Report (2001) of the United Nations Intergovernmental Panel on Climate Change by its advocating a reliance on "carefully constructed scenarios of human behavior and [thus] determine climate projections on the basis of such scenarios" by responding: "… if we can't make accurate predictions about population and development and technology… how can we make a 'carefully-constructed [*sic*] scenario'? What does 'carefully-constructed' mean if we can't make accurate predictions about population and technology and economic and other factors that are essential to the scenario?"[24] He further criticizes the IPCC chart for climate predictions until 2100—which, much as figure 5.5 does—allows for a broad range of futures, ranging from slightly over 1 to 5.8 degrees Celsius. His criticism here is equally blistering: "That is a 400 percent variation. It's fine in academic research. But let's transfer this to the real world." In a point both echoing and emphasizing remarks made here about the frustrations and confusion decision makers have when confronted with such a wide range of future projections, Crichton notes, "In the real world, a 400 percent certainty is so great that nobody acts on it. Ever."[25]

Without doubt, Level 3 uncertainty is far more complex than the schematic mind-sets presented for Levels 1 and 2. Thus, among those who study the human dimensions and impact consequences of climate change, three terms exist to consider when addressing climate change and human impact: resilience, adaptation, and vulnerability. Although still in its fledgling stages, some social scientists have even begun to create indices for human insecurity that measure these factors.[26] Among decision makers, nonetheless, particularly those resistant to accepting the full certainty of climate change as significant impact event, there truly exists only one term: adaptation.

It is for this reason that John Marburger, President G. W. Bush's science advisor, pronounced to a climate scientist at the Goddard Institute for Space Studies (GISS) that "we're really interested in adaptation to climate change."[27] (Note that GISS is a subsidiary of NASA, located on the campus of Columbia University.) The reasons for this interest, as discussed in chapter 7, may primarily be pragmatic economics: Resilience may prove too expensive; and accurate vulnerability measurements may never be possible.

Those of a particular mind-set—particularly those who support the mind-set of a policy maker's stance on the issue of climate change and human impact—may nonetheless intentionally choose to "limit" consideration of a Level 3 range of futures flexibility. One rather glaring example occurred when Philip A. Cooney, chief of staff for the White House Council on Environmental Quality—the office that crafts and promotes administration policies on environmental issues—in 2002 and 2003 altered or removed descriptions of climate research that government scientists and some senior Bush administration officials had already approved. In one instance, despite having a background in economics and no scientific training, he made marginal notations that remarks were "straying from research strategy into speculative findings/musings"[28] (see figure 5.6). In a report stating that "Many scientific observations point to the conclusion that the Earth is undergoing a period of relatively rapid change," Cooney edited the sentence to read: "Many scientific observations *indicate* that the Earth *may be* undergoing a period of relatively rapid change."[29]

President Bush announced his new approach to global warming in February 2002: "When we make decisions, we want to make sure we do so on sound science";[30] notably, these are the same terms Under Secretary Dobriansky consistently used to explain the U.S. isolationist stance on global environmental cooperation. Yet, later that same year, following the Environmental Protection Agency's (EPA's) delivery of a 263-page report to the UN that partially addressed global warming effects, the White House dismissed the report as "bureaucracy"; the White House subsequently interfered in another EPA report so consistently (at one point attempting to insert an excerpt from an American Petroleum Institute partially financed study) that staff members objected, claiming the lack of "accurate scientific consensus"; when the study was later published, the section on climate change was missing.[31]

All of this "bureaucratic" contestation, of course, was not without consequences. Rick S. Piltz, who resigned in March 2005 as a senior associate in the office now known as the Climate Change Science Program, released the documents that Philip Cooney had edited while also declaring, "Each administration has a policy position on climate change …. But I have not seen a situation like the one that has developed under this administration during the past four years, in which politicization by the White House

Figure 5.6

**Changes to U.S. Government Climate Change Research Report. Copyright ©
2003 by the New York Times Co. Reprinted with permission.**

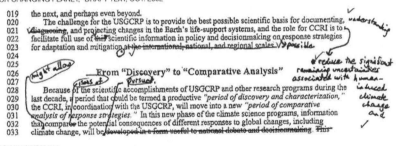

An Editor in the White House

Handwritten revisions and comments by Philip A. Cooney, chief of staff for the White House Council on
Environmental Quality, appear on two draft reports by the Climate Change Science Program and the Subcommittee
on Global Change Research. Mr. Cooney's changes were incorporated into later versions of each document,
shown below with revisions in bold.

"STRATEGIC PLAN FOR THE U.S. CLIMATE CHANGE SCIENCE PROGRAM," DRAFT TEXT, OCT. 2002

PUBLIC REVIEW DRAFT, NOV. 2002
Warming **could** also **lead to changes in the water cycle in polar regions**. Reducing the uncertainties …

FINAL REPORT, JULY 2003
The paragraph does not appear in the final report.

"OUR CHANGING PLANET," DRAFT TEXT, OCT. 2002

FINAL REPORT, 2003
The challenge for the USGCRP is to provide the best possible scientific basis for documenting, **understanding**,
and projecting changes in the Earth's life-support systems, and the role for CCRI is to **reduce the significant
remaining uncertainties associated with human-induced climate change and** facilitate full use of scientific
information in policy and decisionmaking on **possible** response strategies for adaptation and mitigation.

has fed back directly into the science program in such a way as to undermine the credibility and integrity of the program."[32] Philip Cooney, who worked as a "climate team leader" and lobbyist for the American Petroleum Institute prior to coming to work at the White House, resigned his White House position after the release of the documents and took a job with ExxonMobil.

Although this one example may well be extreme, it does serve to illustrate that resistance to Level 3 openness in considering a range of possible futures may well prove to present major setbacks if there is to be any effective policy implemented, at any time, regarding the wide array of potential outcomes, some of them fatal, that link climate change and human impact.

LEVEL 4: WITHOUT A CLUE

Level 4 uncertainty is the most challenging—and most serious—basis for a scenario analysis to occur. Although Courtney, Kirkland, and Viguerie, in their *Harvard Business Review* article on "Strategy under Uncertainty," insist that Level 4 situations are quite rare—and occur often after major technological, macroeconomic, or legislative shock—there can be benefit for those who attempt to shape outcomes so that they can become more recognizable—and approachable—as problems at Levels 3 or 2 of uncertainty. Arguably, the authors display perhaps too much optimism in suggesting that Level 4 scenario situations are transitional. Moreover—particularly apt in considering scenario relevance to climate change and human impact—the authors fail to ever consider the consequence or response necessary in situations in which Level 4 uncertainty becomes *permanent*.

"Global warming is routinely described as a matter of scientific debate—a theory whose validity has yet to be demonstrated"—this characterization, or some form of it, constitutes much of the approach taken by the Bush administration (both father and son).[33] This strikes one as a particularly odd response to long-term, and encroaching, potential vulnerability. Rather, such a characterization is more illustrative of the "do nothing" response, which is the most dangerous response to Level 4 uncertainty.

As regards climate change, Level 4 uncertainty may be more prevalent as a feasible scenario condition than might seem immediately evident. Global warming, for example, is not simply a concept that emerged in the last few decades of the twentieth century. To the contrary, as early as 1859 the physicist John Tyndall, experimenting with what was the first ratio spectrometer, intended to study to the heat-trapping properties of gases. Tyndall soon discovered that gases such as oxygen and nitrogen were "transparent" to visible and infrared radiation; by contrast, CO_2, methane, and water vapor were not. From this discovery, Tyndall suggested that these "nontransparent" gases were largely responsible for the earth's climate—a phenomenon we today might know better as the natural "greenhouse effect." In 1894, the chemist Svante Arrhenius, convinced that humans were influencing the earth's energy balance, posited what the earth would be like if more greenhouses gases were present in the atmosphere—inducing what is termed the "enhanced" greenhouse effect. After a series of extensive calculations, Arrhenius posited that if greenhouse-gas levels increase (with all factors equal), the earth's temperature would rise. Scientists, nonetheless, continued to believe for several decades after these calculations that it was unclear how human-caused (anthropogenic) CO_2 increase would impact the environment—or even if humans were capable of affecting CO_2 levels.[34]

Arrhenius himself predicted that it would take three thousand years of coal burning to double the CO_2 levels in the atmosphere.[35] (Notably, more than 50 percent of electricity energy generated in the United States comes through coal production.)[36] We know that Arrhenius, despite the rigor of calculations that

Figure 5.7
Level 4 Uncertainty: True Ambiguity
No Basis to Forecast the Future. Reprinted by permission
of *Harvard Business Review*. Adapted from "Strategy Under
Uncertainty" by Hugh Courtney, Jane Kirkland, and Pat-
rick Viguerie, 68:3 (November–December 1997), 79–91.
Copyright (c) 1997 by the Harvard Business School Pub-
lishing Corporation; all rights reserved

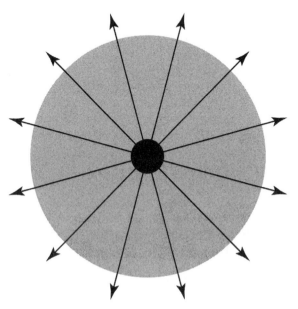

took 14-hour days and nearly a year to complete, illustrated Level 4 uncertainty by erring in his prediction of CO_2 doubling levels by *28 centuries.*

THE NEED FOR FEAR

Ironically, we live in an age not far removed from a time when we lived in constant Level 4 uncertainty. Although the phrase "The Cold War is over" is now no longer the constant refrain that it was in the early 1990s, few seem to remember that the uncertainty that individuals faced in those times, and the strategic policy actions taken, were made in response to an outcome that had never occurred (thermonuclear war). It might prove worthwhile to recall some of the thoughts and ideas from those who attempted to address such "true ambiguity." One of the most powerful precepts to address such uncertainty as climate change is taken from Günther Anders's powerful essay, "Theses for the Atomic Age." One of the most powerful precepts he writes about in the "Atomic Age"—analogous to potential severe outcomes in the "Climate Age"—articulates *The Courage to Fear:*

As a matter of fact, nothing is more deceitful than to say, "We live in the Age of Anxiety anyway." This slogan is not a statement but a tool to prevent us from becoming really afraid, of those who are afraid that once we may produce the fear commensurate to the magnitude of the real danger. On the contrary, we are living in the Age of the Inability to Fear. Our imperative: "Expand the capacity of your imagination," means in concreto: "Increase your capacity of fear." Therefore: don't fear fear, have the courage to be frightened, and to frighten others, too. Frighten thy neighbor as thyself. This fear, of course, must be of a special kind: 1) a fearless fear, since it excludes fearing those who might deride us as cowards, 2) a stirring fear, since it should drive us into the streets instead of under cover, 3) a loving fear, not fear *of* the danger ahead but *for* the generations to come.[37]

We therefore disagree that Level 4 uncertainty is exclusively transitional. To the contrary, Anders's advice from an earlier age of terror seems particularly sage in considering how best to address, consider, and react to climate change and human impact. In a real sense, we have become what Anders would call "Inverted Utopians"—smaller than our true selves, mentally limited, and incapable of true action, response, and production.[38]

Although catastrophic climate change would affect the globe in a disastrous way, scenarios that consider the complex relationship—and uncertain outcomes—from direct impact on human vulnerabilities might compel the serious development of a research agenda that moves beyond the nation-states as the best response mechanism to human impact outcomes as a result of climate change and related vulnerability events. Specifically, if "human security" is the appropriate social (decision-making) unit of analysis, who or what cooperative regime is best suited to provide response? Moreover, what influence or range of responses might be available for such mechanisms to be effective? Finally, given the complex dynamics of this book's focus, we consider it likely that one scenario will not work for all bases of action. Indeed, it well prove worthwhile to consider the options of "nested" scenarios—with multiple options for decision—as an effective policy mechanism for decision makers.

One should therefore not too readily or easily dismiss scenarios that seem "extreme." Extreme outcomes, in the context of climate change, cannot be ruled out. When mollusks are found growing only several hundred miles from the Pole, or entire species are in permanent migratory pattern change, or butterflies throughout the Northern Hemisphere are shifting ranges northward by up to 150 miles, or amphibians such as the Golden Toad of Costa Rica are eliminated because they can only survive at specific altitudes (and at specific temperatures in ecological balance), or predatory insects or plants begin to invade ecosystems or undermine biodiversity, or even polar bears increasingly drown because of the loss of surface ice, then all is not well. One has, at such point, reached a "tipping point"—commonly described through the visual anecdote of leaning over in a canoe to the extent that one is no longer "rocking the boat" but flipping it, and from which there can be no recovery.

We cannot know with any precision if we have reached that tipping point. But we return to the argument presented earlier in this work: We are almost always forced to change, not by strategic design, but by the impact of the unknown, which changes the rules of the game. We may well be entering the realm of the ambiguous unknown.

There are reasons to be afraid.

POLICY IMPLICATIONS AND APPROPRIATE RESPONSES

With regard to the complex relationship between climate change and human impact, scenarios could prove useful in determining policy and action response. By developing such scenarios, the interrelationship between that which is predetermined and that which is uncertain may be equally open to interpretation and changing factors. Pierre Wack offers several thoughts with respect to the use of scenarios as tools:

> I have found that scenarios can effectively organize a variety of seemingly unrelated economic, technological, competitive, political, and societal information and translate it into a framework for judgment—in a way that no model could do.... Decision scenarios describe different worlds, not just different outcomes in the same world....You can test the value of scenarios by asking two questions: (1) What do they leave out? In five to ten years...[decision makers] must not be able to say that the scenarios did not warn of important events that subsequently happened. (2) Do they lead to action? If scenarios do not push managers to do something other than that indicated by past experience, they are nothing more than interesting speculations.[39]

We are experiencing a world of dynamic change where even the most mind-numbing, dramatic events do not impress us for long. Yet any good strategist and planner must be able to help the nation's leaders see more clearly the different futures that may occur. To operate in an uncertain world, we need to *reperceive*—to question our assumptions about how the world works, so that we see the world more clearly. The purpose of this reperception is to help make better decisions about the future.

With respect to specific improvement of scenario-based studies of climate change and human/national security, we might offer four recommendations:

A. Consult with experts in security studies in addition to climate and environmental experts. As noted previously, the relationships between the environmental and traditional national security concerns have been a source of debate. Although a consensus of opinion has not been reached, much has been done to productively move the discussion forward. Since the 1990s, the topics of study have evolved from efforts to identify which aspects of the environment could usefully and practicably be applied to security questions[40] to more complex considerations of how environmental factors contribute to some conflicts, but not to others.[41] Indeed, given the increasing number of journal articles and professional conferences, *environmental security* can be recognized as an important subdiscipline of security studies, if not a discipline in its own right. Products of this work include conceptual clarifications and models of structural relationships among factors that broadly define a security context. In chapter 3, for example, we distinguished between *threats,* which are readily identifiable and commonly understandable in conventional terms of security, and *vulnerabilities,* which expose long-term stability to risk without a direct threat. Vulnerabilities are difficult to directly perceive but are evident by indexes such as hunger, disease, and poverty. An example of the latter is Thomas Homer-Dixon's model of "Environmental Scarcity and Violence."[42] By employing these analytical tools in the scenario development and assessment processes, climate change futures studies can better engage the needs of security decision makers.

B. Identify or develop frameworks for inferring reasons for action at the regional, nation-state, and super-state scale of organizations. The structural mappings, such as by those Homer-Dixon, are useful for understanding the relationships among elements that comprise a model of an environmental security. However, such mappings do not, by themselves, allow for an understanding of how specific sequences of events and actions may transpire over time to destabilize a security situation. There is, therefore, also the need to understand the dynamics of change. The approaches described by Clayton Roberts and Kenneth Burke have been mentioned as possible ways to infer the logic of choice and can be used for scenario development.[43] Other frameworks used in literary analysis, history, anthropology, and sociology exist and should be recognized, adapted, and applied to help structure representations of the future. Of particular need are frameworks that link the relationships between societies and their environments. Because our relationship to our environment is a complex fusion of biological needs, religious beliefs, social values, and economic goals, employing any framework to better understand how people act with regard to "nature" may be controversial. Relative to our own point of view, we may judge the actions of others as misguided or morally wrong, but the reasons for such actions exist nonetheless. As John Watkins has conscientiously noted, "To regard the doings of another as essentially unexplainable is to regard him [or her] as not quite human."[44] Although we may not—or should not—excuse anyone or any organization for any unfortunate consequences that follow from poor decisions, efforts to make sense of reasons for action help us to better grasp the workings of our social world.

C. Use the identified framework(s) to describe case studies of past actions in times of environmental stress. Although the future may not be like the past, historical precedent can serve as a guide to understand and anticipate some of tomorrow's actions. Short of a systematic library, there are a growing number of historical accounts that could contribute to a collective knowledge base. As noted in chapter 2, LeBlanc's work (cited by Schwartz and Randall) is a useful start. Another example is Brian Fagan's *Floods, Famines, and Emperors,* which explores how some cultures have adapted to climatic change and others have not.[45] On a similar theme is Mike Davis's *Late Victorian Holocaust,* which is "a political ecology of famine,"[46] and examines the interactions between climatic and economic processes, interactions that have resulted in the kinds of social upheavals that may be reminiscent of future security concerns.

D. Borrow a page from military theory and distinguish among strategic, operational, and tactical thinking and develop scenarios specific to each kind of decision making in addition to synthetic visions of the future. The decisions to adopt a defensive environmental posture, to break a longstanding treaty with Mexico, and to administer border control through the Department of Defense may be related within a broad arc of geopolitical understanding, but each represents a different level of engagement with the world. On the one hand, one strength a of scenario that spans multiple levels of decision making is that it can be invaluable for understanding the breadth of choices that might need to be made. But on the other hand, such a scenario may also suffer from a lack of depth and detail with regard to any single choice. This lack may impede discussions about the future by both those who will help to inform the decision and those who will be affected by it. The three decisions mentioned here could—and perhaps should—all be of interest to the office of the President's National Security Advisor, to the State Department, and to the Department of Defense's Northern Command, but it is not necessarily clear that each organization would be equally interested in the implications of all three decisions. Furthermore, each of these organizations operates in its own political context, has its own mission, and takes different kinds of actions. As such, it could readily be imagined that each organization would, if given the opportunity, request its own unique set of further elaborations about the scenario in order to better evaluate the situation, assess its options, and plan its actions. Developing environmental scenarios (or any kind of scenarios) that are focused to strategic, operational, or tactical decisions will help deepen the discussion of policy options at each level.

The security threats posed by abrupt climate change will potentially affect every nation and every person on the planet in negative ways. Those nations that are better prepared may, in the long term, be better off, but that is not to say there will be winners and losers. Today, new frameworks to assess the interconnectedness of national and global security are being developed and should be used to understand common concerns and common solutions. That climate change is being discussed at all under the rubric of "national security" indicates that the kinds of threats that are recognized have greatly increased since the end of the Cold War. Reflecting this new situation, the range of specialists who can provide useful, if not much needed, insight has necessarily expanded. Today, seminars on security studies require a larger table.

These contextual differences lead to new opportunities and requirements for scenario development. Because the discussion can be open, assumptions about the future can be more explicitly stated. Because the discussion should seek, explore, and chart new relationships between society and the environment, the assumptions about the future should be more explicitly stated. And because the discussion must include people from multiple disciplines and diverse backgrounds who must talk to each other, the assumptions about the future must be more explicitly stated. This is not to say that these focused scenarios should be considered as substitutes to synthetic scenarios, such as the one written by Schwartz and Randall. Both kinds of scenarios are needed to help decision makers better consider the dynamics of change and better meet the demands of possible future situations.

Perhaps one way to think about this is to obvert George Santayana's famous saying about learning from history by changing our perception of things that are yet to come, by suggesting that "those who do not learn from the future are destined to make mistakes in it." To be able to understand that future, we must have a "mental map" flexible enough to consider plausible alternatives and possibilities we might not otherwise consider. Given how the scientific data are almost insurmountable about the impact of climate change on humanity now—and not at some indefinite time in a murkily foreseen future—it remains disturbing how slow policy makers have been in their responses to affect change. The United States, ultimately, with its largely noncooperative stance on binding climate protocols, is the most important actor in this process, if only because it is also the largest greenhouse gas emitter. Indeed, with 4.5 percent of the world's population yet responsible for 34 percent of greenhouse gas emissions, the United States is an actor in effecting change that cannot be ignored.[47] As this chapter illustrates, nonetheless—even with the necessary imposition of the "fear factor"—affecting that change, overcoming mind-sets, and implementing correct policy actions will not be easy, and may not be possible.

To return to the concept of "manufactured uncertainties," discussed in chapter 3, both a conundrum and a real paradox emerge when increased knowledge fuels the production of ever more increased uncertainty.[48] Added to this is the recognition that in furthering the development of humanity through the use of ever-improving technologies (and the side effects of such technology as pollution and greenhouse gas emissions), humanity itself may well feel a negative impact engendered by such "improvement." Thus, our continuing emphasis throughout this book centers on climate change and human impact.

The Kyoto Protocol expires in 2012. The greenhouse emission "cuts" that the protocol mandates, by any standard, do not approach levels necessary to bring down worldwide emissions. Indeed, even if every country in the world (including the United States) were to fulfill Kyoto requirements, CO_2 concentrations in the atmosphere would still be headed to 500 parts per million.[49] The highest previous CO_2 concentrations occurred roughly 325,000 years ago.

In the end, we can be certain of one thing: The future will be challenging.

NOTES

1. Hugh Courtney, Jane Kirkland, and Patrick Viguerie, "Strategy Under Uncertainty," *Harvard Business Review,* November–December 1997, pp. 79–91.

2. Ibid.

3. Ibid.

4. Robert C. Whitten, "In My View: Greenhouse Gases and Global Warming," *Naval War College Review* 56, no. 1 (Winter 2003) pp. 142–45, http://www.nwc.navy.mil/press/Review/2003/Winter/imv-w03.htm.

5. Elizabeth Kolbert, "Annals of Science: The Climate of Man—I," *New Yorker,* April 25, 2005, p. 63.

6. For further background on late-twentieth-century scientific thought on "accurate" thermometers in space, see http://science.nasa.gov/newhome/headlines/essd06oct97_1.htm. For two contradictory media reports of the studies subsequently referenced here in *Science,* see Steven Milloy, "Global Warming Doubt Dispelled? Not Really," August 19, 2005, FoxNews.com. http://www.foxnews.com/story/0,2933,166150,00.html; Dan Vergano, "Scientists Find Errors in Global Warming Data: Satellites Gave Temps for Nights, Not Daytimes," *USA Today,* August 12, 2005, p. A3.

7. Steven C. Sherwood, John R. Lanzante, and Cathryn L. Meyer, "Radiosonde Daytime Biases and Late-20th Century Warming," *Science* 309 (September 2, 2005): pp. 1556–59.

8. Ker Than, "Key Claim against Global Warming Evaporates: Satellite and Weather Balloon Data Based on Faulty Analyses, Studies Find," http://msnbc.msn.com/id/891703/print1/displaymode/1098.

9. Carl A. Mears and Frank J. Wentz, "The Effect of Diurnal Correction on Satellite-Derived Lower Tropospheric Temperature," *Science* 309 (September 2, 2005): pp. 1548–51.

10. Than; B. D. Santer, T. M. L. Wigley, C. Mears, F. J. Wentz, S. A. Klein, D. J. Seidel, K. E. Taylor, P. W. Thorne, M. F. Wehner, P. J. Gleckler, J. S. Boyle, W. D. Collins, K. W. Dixon, C. Doutriaux, M. Free, Q. Fu, J. E. Hansen, G. S. Jones, R. Ruedy, T. R. Karl, J. R. Lanzante, G. A. Meehl, V. Ramaswamy, G. Russell, G. A. Schmidt, "Amplification of Surface Temperature Trends and Variability in the Tropical Atmosphere," *Science* 309 (September 2, 2005): 1551–56.

11. Whitten, "In My View: Greenhouse Gases and Global Warming," *Naval War College Review.*

12. Elizabeth Kolbert, *Field Notes from a Catastrophe: Man, Nature, and Climate Change* (New York: Bloomsbury, 2006), p. 157.

13. Ibid., 170.

14. Michael Shellenberger and Ted Nordhaus, *The Death of Environmentalism: Global Warming Politics in a Post-Environmental World,* http://www.thebreakthrough.org/images/Death_of_Environmentalism.pdf (May 27, 2005), pp. 6–7.

15. Ibid., pp. 14–15.

16. Kolbert, *Field Notes from a Catastrophe,* p. 160.

17. Ibid., pp. 149, 150.

18. Quoted in Chris Mooney, "Some Like It Hot," *Mother Jones,* May/June 2005, pp. xxx–xxx.

19. Paula Dobriansky, "Technology Solution to U.S. International Climate Change Issues," Capitol Hill Hearing Testimony to the Senate Foreign Relations Subcommittee on International Economic Policy, Export and Trade Promotion, November 14, 2005.

20. Juliet Eliperin, "U.S. Won't Join in Binding Climate Talks: Administration Agrees to Separate Dialogue," *Washington Post,* December 10, 2005.

21. Courtney, Kirkland, and Viguerie, 77.

22. Address to the American Enterprise Institute, quoted in Mooney, "Some Like It Hot," p. 1.

23. Michael Crichton, "The Impossibility of Prediction," speech to the National Press Club, Washington, DC, January 25, 2005, http://www.crichton-offical.com/speeches/npc-speech.html.

24. Ibid.

25. Ibid.

26. See, for example, AVISO, "The Index of Human Security," *Aviso,* January 2000, http://www.gechs.org./aviso/avisoenglish/six_lg.shtml; Gary King and Christopher J. L. Murray, "Rethinking Human Security," *Political Science Quarterly,* 116: 2001–2002. In the interest of transparency, the authors note that P. H. Liotta is an associate of the Global Environmental Change and Human Security (GECHS) project. For a critique of the perceived deficiencies of these indices—primarily questioning how these efforts attempt to measure development issues rather than focus on violence, and are not regularly updated—see *Human Security Report 2005: War and Peace in the 21st Century* (Oxford, UK: Oxford University Press, 2006), p. 90.

27. Kolbert, *Field Notes from a Catastrophe,* p. 108. Marburger, notably, is previously referred to in this chapter as the leader of the assessment regarding the "strangelet scenario" and the Brookhaven RHIC particle accelerator.

28. Andrew C. Revkin, "Bush Aide Softened Greenhouse Gas Links to Global Warming," *New York Times,* June 8, 2005, http://nytimes.com/2005/06/08/politics/08climate.html.

29. Kolbert, *Field Notes from a Catastrophe,* p. 164.

30. Extracted from http://www.nytimes.com/imagepages/2005/06/07/politics/20050608_climategraph.html.

31. Kolbert, *Field Notes from a Catastrophe*, p. 163.

32. Ibid., 163–64.

33. Courtney, Kirkland, and Viguerie, pp. 78, 79.

34. Kolbert, "Annals of Science: The Climate of Man—I," p. 71.

35. Extracted and condensed from ibid., pp. 65–66.

36. Ibid., p. 66.

37. Elizabeth Kolbert, "Annals of Science: The Climate of Man—III," *New Yorker,* May 9, 2005, p. 56.

38. Günther Anders, "Theses for the Atomic Age," *Massachusetts Review* 3 (Spring 1962): p. 498.

39. Ibid., p. 496.

40. Pierre Wack, "Scenario: Shooting the Rapids," *Harvard Business Review* 63, no. 6 (November–December, 1985): 146–47.

41. T. F. Homer-Dixon, "On the Threshold? Environmental Changes as Causes of Acute Conflict," *International Security* 16: (1991): pp. 76–116.

42. G. Dabelko, "The Environmental Factor," *Wilson Quarterly* 23, no. 4 (1999): pp. 14–19.

43. T. F. Homer-Dixon, *Environment, Scarcity, and Violence* (Princeton, NJ: Princeton University Press, 1999).

44. A. W. Shearer, "Applying Burke's Dramatic Pentad to Scenarios," *Futures* 36, no. 8 (2004): 823–35.

45. J. Watkins, "Imperfect Rationality," in *Explanation in the Behavioral Sciences,* ed. R. Borger and F. Cioffi (Cambridge, UK: Cambridge University Press, 1970), pp. 167–217; this note p. 216.

46. Brian Fagan, *Floods, Famines, and Emperors: El Niño and the Fate of Civilization* (New York: Basic Books, 1999).

47. Mike Davis, *Late Victorian Holocausts: El Niño Famines and the Making of the Third World* (New York: Verso, 2001).

48. As defined by Annex I of the Kyoto Protocol. Common estimates among U.S. policymakers who acknowledge the impact of the United States on affecting global warming conditions is that the U.S. accounts for roughly 25 percent of the world's greenhouse gas emissions. See, for example, Kolbert, *Field Notes from a Catastrophe,* pp. 169–70.

49. Ulrich Beck and Johannes Willms, "Global Risk Society," in *Conversations with Ulrich Beck,* trans. Michael Pollak (London: Polity, 2004), p. 127.

— 6 —

Rethinking the Unthinkable: Comprehending Climate Change Futures

The future itself is a "work of art" until it is actually upon us.

Kenneth Burke

When considering the use of scenarios to understand the potential consequences of global warming and climate change, it is perhaps ironic that the widespread use of scenarios as a means to improve long-range thinking is often attributed to the strategic planning successes of the multinational oil company Royal Dutch/Shell.[1] To be sure, all of us who help develop scenarios and the many more who use them owe intellectual debts to those at Shell along with corporate planners and MBA-wielding consultants from around the world. Their substantial methodological advances have helped in the daunting task of envisioning the future.

In this chapter, we expand upon the prevalent thinking about climate change scenarios by drawing upon analytic techniques from the broader futures-oriented literature.[2] Any exercise of long-term strategic thinking is difficult due to the inherent unknowns of what tomorrow might bring. Climate futures are especially demanding due to complexities of the science, limited records of the past, and incomplete knowledge about the influences of human activities on natural systems. Furthermore, when it comes to matters of security, there is an uneasy social calculus that often involves trading one kind of protection to gain another. Our position is that these difficulties do not give us an excuse to not think about these challenges. Quite the opposite, it is because they are so difficult that we must devote concerted and disciplined effort to understand how we might proceed to address them. As this generation thinks about its unthinkable problem—to again borrow Herman Kahn's phase—science plays a crucial role by allowing us to quantify relationships among geophysical, biological, and social systems. However, though the impacts of human activities might be precisely and accurately assessed, understanding the motivations for these actions is more complex. If we are to understand how we might slow climate change or adapt to it, we also need frameworks to represent motives, means, and opportunities.

We want to call particular attention to some methods of scenario development that seem especially important for understanding the relationships between changing ecological contexts and the spectrum of security concerns. We need ways that describe how human activities, which might both contribute to and respond to climate change, can be adequately represented for the purpose of presenting plausible scenario storylines. Capturing the logic, or perhaps more correctly, the rationale, of all the things we individually and collectively do is very difficult. The damning phrase of a bad movie is often, "But the character would never do *that*." It could be that we would believe the character would, indeed, do *that* if we understood more about the situational context, more about the way the character has behaved in the past, or more about what the character anticipated.[3] Recall that in their scenario on climate change and security, Schwartz and Randall made the assertion that "every time there is a choice between starving and raiding, humans raid."[4] We argued that this generalization seems overly simplified given the current predominance of the nation-state as the geopolitical unit of measure and the relationships between individuals and nation-states. But that point of critique begs a question: If the stated assumption about human activities is insufficient—or, at least, seemingly unsupported—what framework is better suited to take its place? More directly, for the purpose of developing scenarios to engage climate change and security, how might we frame action choices? If those with responsibility to take steps for society's benefit cannot understand the "story logic" of the scenario, then there is the potential that the scenario will not be considered at all. In the end, the decision maker may disagree with certain premises of the story logic, but if the logic is transparent, then there is the opportunity for a conversation; if the logic is opaque, then there is a greater chance for out-of-hand dismissal.

It cannot be overemphasized that there are no "future facts" relating to human enterprise; there are only assumptions. And the handling of assumptions requires a few preliminary comments. Because the assumptions are provisional (at least until our future prospects become historical records), we should always be very careful in specifying our explanations and reasons for change. Moreover, we must be aware of the cumulative impression of individual assumptions that contribute to a larger scenario. Novelist and essayist E. M. Forster distinguished two kinds of readers:[5] The curious reader is one who simply moves from fact to fact, continually asking (and only asking), "and then?" By contrast, the intelligent reader approaches each fact from two vantages: first, in isolation and second, in light of everything that has already been recounted. The former perspective identifies just what the fact *is*; the latter perspective attempts to understand what the fact might *mean*. Most importantly, the intelligent reader knows that part of the comprehension must be suspended because the true significance of the fact may not be revealed until later in the story after other cross correspondences have been exposed. Scenario users must likewise be intelligent readers of assumptions.

ASPECTS OF THE FUTURE

Making any decision involves an assessment of the future. In part, we wonder how our actions may change our own the current situation and influence the situations of others—for better and for worse. Also in part, we are concerned with how things beyond our own control will affect our prospects for success. In all cases, whether our thoughts are about the potential outcomes of personal agency or the potential consequences of contextual dynamics, the discussion of the future rests on assumptions rather than facts. Building such a logically contingent edifice requires extreme care. It is therefore critical that we choose the assumptions that provide our metaphoric pillars and plinths carefully. As Bertrand de Jouvenel noted, "What is of vital importance for the progress of this art of conjecture is that an assertion of the future should be accompanied by the intellectual scaffolding which supports it: this 'construct' must be transparent and articulated, and subject to criticism."[6] Various ways for constructing such a scaffold of thought about the future have been devised.[7]

In a world in which the range of possible futures can seem infinite, it can be operationally beneficial to explicitly identify which dimensions of change are being considered in a given scenario and what is assumed to be held constant. Perhaps more importantly, scenarios are not just stories about the future, but stories that are envisioned and shared in order to think about the consequences of decisions that must be made in the present. Classifying different kinds of change can be a useful first step in identifying appropriate preventative measures, recovery options, and restoration potentials. Although sweeping narratives of tomorrow can help to bring a shared vision to a large number of people—which is of no small benefit—decision makers often need a more complete understanding of the details embedded in the assumptions in order to act. As such, not all stories or all kinds of change are salient to all groups. For the purposes of thinking about relationships between climate change and security, we believe there is practical benefit in using a framework proposed by Kenneth E. Boulding. It divides kinds of potential change into four groups: planets, plagues, plants, and plays.[8]

Planets: Planets are elements of systems with highly stable and recognized parameters. This stability allows these aspects of the future, such as the timing of a solar eclipse and the location where its effects will be most prominent, to be predicted with a high level of confidence. Admittedly, few systems behave with the regularity of celestial mechanics.

Plagues: Plagues are elements of the natural world such as earthquakes, hurricanes, tornadoes, tsunamis, floods, and droughts that can cause devastating impacts to society. Unlike the regularities evidenced with planets which allow prediction of events far in advance, it is difficult to anticipate the timing, frequency, location, and intensity of plagues. We can however, sometimes identify the structural conditions that make these events relatively more likely and watch for indicators of near-future occurrences. For example, we know the path of the San Andreas geological fault that runs through California. By monitoring of seismic activity, it is possible to better anticipate significant events. Sometimes the structural conditions that allow for plagues have a discernable periodic dimension, which, in and of itself, allows a very general notion of what might occur in the future. For example, we know that warmer surface water temperatures in the Atlantic fuel a "hurricane season."

Plants: Plants are elements that begin with a seed and, once set in the earth, sprout, develop, and eventually die. The distinguishing characteristic of these parts of the future is that they contain an image of the expected result before the process of becoming begins, and come what may, this image is persistent. While initial conditions and developing external forces may influence the final form of the plant—that is, how large it will become, how long it will survive, how much fruit it will yield—it still grows according to its preestablished set of rules. This persistent image also means that plants hold specific kinds of potentials and not others. Thus, a sunflower seed, given appropriate nourishment and protection, will become a sunflower; it will not become an oak tree.

Plays: Plays are the real-world dramas of our lives. The defining characteristic of this aspect of the future is an emergent quality that arises through the continuing interactions of many individuals and groups with each other. Hence, unlike the persistent image of the future which assiduously directs the growth of a plant to a predefined goal, plays are open-ended. While many plays are similar to each other—like being part of the same genre—the conclusion cannot be predicted.

Like all frameworks, Boulding's four-part scheme can be applied to many kinds of interests and problems. Relative to climate change and human security, our concerns are the sources of uncertainty inherent to each kind of change and, in particular, the vulnerabilities that result.

THE FUTURE AND CLIMATE CHANGE

Planets

The solar system, which serves as the basis for the planet metaphor of the future, is relatively stable over very long periods of time. Indeed, it is so stable that ancient astronomers were able to anticipate the paths of stars and planets. However, though the calculations that allow us to fix the future locations of celestial bodies are deterministic, there are degrees of cyclical variability within the system that complicate the math. Although seemingly slight in comparison to the scale of the universe, the earth's orbit around the sun displays three kinds of variability, which affect the amount of energy that can be received and the relative severity of summer and winter.[9] These variations are sometimes referred to as the (Milutin) Milanković Cycles, named after the Serbian mathematician and engineer who advanced their importance on the development and recession of past ice ages.

The first variability is axial tilt; that is, the angle of the earth relative to its orbital plane. Today the tilt of the axis is 23.4 degrees—just less than its average mark—but over a 41,000 year cycle it varies between 21.5 and 24.5 degrees. It is this tilt that gives us the seasons. The hemisphere that is tilted toward the sun experiences summer, having longer days and receiving sunlight at a more direct angle. Of note for global climate variability, energy received in the northern hemisphere is more influential because it has more land mass and can therefore potentially absorb heat faster than the ocean-dominated southern hemisphere. As the earth orbits, the other hemisphere transitions from winter to summer as it comes to be tilted toward the sun. When the axial tilt decreases (say, from 23 degrees to 22 degrees), higher latitudes receive less solar radiation in the summer months and, consequently, less ice is melted. Over time, a feedback loop, which amplifies the effects of shorter summers, begins: The persistent high albedo (a measure of reflectivity) of unmelted snow and ice bounces more of the sun's energy back into space. The redirection of sunlight increases the cooling of the earth which, in turn allows for even more snow and subsequently more cooling. Conversely, when the axial tilt increases, higher latitudes receive more solar radiation in the summer months. The more summertime energy, the more ice is melted. In this case a different feedback mechanism comes into operation: The melting snow reveals rocks and soil (and eventually plants) that lower the earth's albedo, making it possible for the planet to absorb more heat. And as more energy is absorbed by the surface, the temperature rises, melting more snow and allowing for more rocks and soil (and plants) to absorb more heat from solar radiation. And as temperatures rise higher in high latitude areas, it is also possible that methane, which is locked in frozen soils, is released, further increasing the atmosphere's capacity to store heat.

The second variability is the eccentricity of the Earth's orbit. As it moves around the sun, our planet follows an ellipse that, over time, changes in shape. The most influential part of this variation occurs on a cycle of 413,000 years, but other components vary roughly every 100,000 years. The point at which the earth is closest to the sun—and thereby receives the most intense solar radiation—is called the perihelion; and the point at which the earth is farthest from the sun is called the aphelion. When the orbit is nearly circular, as it is now, the intensity of solar radiation at the perihelion is about 6 percent stronger than when the earth is at the aphelion. But when the orbit is more elliptical, then the earth's path takes it both closer to and farther away from the sun. Then the difference in solar radiation between perihelion and aphelion is over 20 percent.

The third variability is the "precession of the equinoxes." The earth's axis not only varies in its tilt; it also wobbles like a spinning top on a 22,000 year cycle. As a result of the wobble, the time between perihelion events is just slightly more than one Georgian calendar year. So, the season during which the earth is closest to the sun changes over time. In principle, the hemisphere that experiences summer at perihelion and winter at aphelion is subject to more severe seasons than the opposite hemisphere. Currently, the earth reaches its perihelion in early January, a situation that contributes somewhat to relatively milder winters in the northern hemisphere. But when the perihelion occurs in early July, as it did 11,000 years ago, the intensity of the summer in the northern hemisphere is significantly increased.

Because humans have no influence over celestial movements, the astronomical impacts on the climate can be readily classified as events rather than actions. Yet though mapping these periodicities is straightforward, there are two complicating issues. First, the duration of these cycles presents a mental challenge for the scenario writer and user. It is too difficult for most people—arguably for anyone—to imagine a story of human endeavor and societal change that spans such a gulf of time. The time horizon for strategic planning scenarios is sometimes as far ahead as 100 or 200 years,[10] but 20 or 25 years—the time it takes for a generation to come of age—is more common. Many planning activities, such as the U.S. military's quadrennial review or the old Soviet five-year plans, assume far shorter prognostication. The further one peers into the future, the more difficult it is to imagine oneself or one's peers. At some stage of tomorrow, the current generation simply ceases to be in the image of the future; then, arguably, the salience of future situations fades, and the ability to understand motivations weakens. What will the great-great-great-great-great grandchildren of today's parents need and want? To get a sense of how difficult that question can be to answer, think for a moment how much civilization has changed in the past 200 years alone. Nonetheless, since variations of the earth's orbit can provide the conditions for an ice age to develop, they present a fundamental vulnerability for life on this planet. And so despite the conceptual difficulties in devising and using scenarios for this factor of climate change, the factor cannot be dismissed.

The second issue that complicates the development and interpretation of this class of scenarios is that the astronomical explanation for glacial advance and retreat does not include the influence of human-induced global warming. Based on the geologic record, it can be stated that in the Pleistocene, interglacial periods—the time between ice ages—lasted roughly 10,000 years on average and generally not longer than 12,000 years. The Holocene Epoch, which is the current interglacial period, has already lasted about 10,000 (radiocarbon) years, and so one could argue based on past events that we are due, if not overdue, for a cooling period.[11] And in a reappraisal of a seminal computer simulation that corroborated the influence of Milanković Cycles on ice ages, William F. Ruddiman's research suggests, in contrast, that the earth should have started drifting toward a new stage of glaciation 6,000 to 5,000 years ago.[12] What delayed the process, he argues, was the release of greenhouse gases that began with the earliest agricultural practices. Some 8,000 years ago carbon dioxide (CO_2) was released by the burning of forests to create fields. In addition, domesticated animals contributed to increasing the concentration of the more powerful greenhouse gas methane. Some 5,000 years ago, the advent of irrigated rice fields (artificial wetlands) began to release still more methane. According to Ruddiman, by the start of the industrial era, human-spurred emissions of CO_2 and methane had "caused increases in the atmospheric concentrations of both gases equivalent to about half of the natural range of variation that had occurred previously."[13] He qualifies that the net result of preindustrial man plus nature was still within overall range of natural variability, but it was near the top of that range. Although we are all now well aware of the influence of the industrial age on the atmosphere, Ruddiman's thesis, which is not without its skeptics, leads to the conclusion that even the most basic practices of farming, which allow for widely dispersed small villages, can affect climate. Given this evidence, will humanity ever see the readvance of ice sheets in the northern hemisphere even if we create less polluting manufacturing and agricultural processes? If what is suggested by Ruddiman's research on the past is true, do we need to bother thinking about astronomically induced climate changes?

One scenario that draws attention to the consequences of planet-type concerns in the not-too-too-distant future was outlined by the noted science fiction writer Arthur C. Clarke. Writing for a special edition of *AsiaWeek* in 1999 on the then-coming new millennium, he made the prediction that by the year 2090, we would start to burn coal for the express purpose of fending off an ice age.[14] His article was short, and he supplied, rather than a tightly woven narrative, a list of brief statements. The relevant actions leading up to our newly pressing reason to release CO_2 into the atmosphere will be the invention in 2002 of low-temperature nuclear reactors and in 2010 of the quantum generators that tap space energy. And perhaps even more significantly, relative to Ruddiman's thesis about agriculture as a source for greenhouse gases, by 2045 we are mining CO_2 from the air to boost food synthesis.

Obviously, Clarke, who has previously displayed what might be called a knack for anticipating future technological developments, missed a few marks in this projection of technological advancement. For starters, the replacement of carbon-based fuels by cold fusion has not transpired. Still, even discounting any patent-filing details of the time line, Clarke's conjecture is well worth considering in the context of long-range strategic thinking and in the awareness of vulnerabilities. We know from science (not science fiction) that by using fossil fuels in such high amounts as we do now, we are currently increasing the concentration of greenhouse gases in the atmosphere. This intense use contributes to global warming and climate instability. In turn, this instability may create a condition of vulnerability relative to the continued needs to grow crops, manage the spread of disease, protect life and property from disasters, and so on. And so, at first reading, Clarke's outline of possible future events is very optimistic in that it describes a solution to the problem of global warming by the creation of abundant, safe, reliable, nonpolluting energy sources. But this very success exposes—actually re-exposes—life on the planet to the vulnerability of the Milanković Cycles and a new ice age. This possibility suggests that we should be conserving fossil fuels not only to lessen the near-term impacts of climate change, but also because these fuels are a resource that, if released during a different planetary ecological context, could contribute to the survival of future societies.

Clarke's image of the future will no doubt leave some with a sense that this scenario is too far "out there," either for its relatively long time horizon or its technological speculations. However, drawing out the consequences of this plot line leads to two general points about scenario creation and use. First, our perception and prioritization of vulnerabilities change over time. Second, too rarely do individuals or governments ask what happens after success. There should be no mistake: The innovation of clean and renewable energy would be a "problem" that any person or government would be happy to have. But if we do have the need to think beyond the usual 20-year planning horizon, then we must also be willing to accept the responsibility to think beyond what future generations will see as short-term gains.

Plagues

The uncertainties associated with plagues are the location, timing, frequency, intensity, and duration of these devastating localized events. Like planets, plagues result from the forces of nature and can therefore be considered events, rather than actions; however, unlike our inability to influence the gravitational forces that move stars, there are some actions that humans can take to affect the probability of plague occurrences and the vulnerability to plague impacts.

Contributing to climate change, one kind of plague that is beyond our influence is volcanic explosion. An eruption sends dust and sulfur dioxide into the atmosphere. The heavier dust and ash particles soon settle or are returned to the surface with rain, but the sulfur dioxide converts to sulfate aerosols, which can stay aloft for several years. This layer of particles intercepts and reflects solar radiation, lowering surface temperatures; however, they can also cause temperatures in the atmosphere to rise. Given the circulation patterns of the upper atmosphere, climate effects are usually limited to the hemisphere in which the eruption occurred, although events in the tropics can affect global climate. Although the cooling effects are relatively short term, they can be significant. For example, the 1991 eruption of Mount Pinatubo in the Philippines lowered the global average surface temperature by 0.3 degrees Celsius while raising the temperature in the tropical lower stratosphere 2–3 degrees Celsius.[15] The most extreme example in *recorded* history was likely the eruption of the Tambora volcano on Indonesian island of Sumbawa in 1815, which resulted in what was known in the New England states as the "Year without a Summer" (1816).[16] Snow fell in June and crops in the United States and Canada were destroyed by frosts throughout the growing season, and in Europe cold and wet weather produced food shortages. The disruptions of rainfall patterns also affected China and southeast Asia. And as noted in chapter 3, a far more violent event may have occurred 73,000 years ago in what is today Sumatra. Then ash circled the earth for several years, photosynthesis essentially stopped, and the precursors to what is today the human race amounted to only several thousand survivors worldwide.[17]

Other kinds of plague may follow from climate change. These include more severe river and coastal flooding,[18] water shortages and drought,[19] extreme heat,[20] the spread of disease,[21] and the loss of biodiversity.[22] Here, we must recognize that not only will we be vulnerable to these situations, but our current actions that contribute to global warming contribute, in turn, to the conditions that will allow these events to occur with increasing severity. Although it is not possible to say that CO_2 emissions caused, for example, Hurricane Katrina, which devastated the Gulf Coast in 2005, it can be said that we have warmed the planet, and warmer sea surface temperatures fuel hurricanes.

A trusted response to plague events has been insurance: the purchase of a financial hedge against the possibility of undesirable occurrences. Although insurance does nothing to prevent the events from occurring or to mitigate the impacts as they are unfolding, it does provide monetary compensation after the fact, which can allow the repair or replacement of damaged or lost materials. Since (at least) the fifteenth century, insurance helped make commerce work,[23] and today many depend on it to protect assets for themselves and for their families' well-being. An insurance transaction is based on the ability of the insurer to establish a measure of *risk*—that is, the probability of loss or harm from a natural or artificial process. This understanding of risk brings together the notions of threat and vulnerability. In the context of climate change, the threat is some plague-like event in the environment that has been brought about by whatever means. Threat, in this sense, is a hazard. The potential impact of the threat is, in part, a matter of the cash value of the loss and also, in part, the capacity of the person, thing, or place to withstand the assault. A valuable object placed in a setting that offers good protection is less at risk than the same object placed in a setting that offers little or no protection. Event probabilities are based on the patterns of occurrences in the past. So, as long as the climate was stable over a long period of time, the probabilities were also stable. But climate change will undermine this empirical basis and make premiums more difficult to price. Moreover, what might be lost is becoming more expensive to replace. Over time, the cost of rebuilding cities following catastrophic events such as floods and hurricanes has grown not only by inflation, but by the greater concentration of people and resources in cities and by the greater penetration of higher technology into urban infrastructures.[24] Even worse, the loss of life associated with this kind of disaster is not only an immediate humanitarian loss, but also presents a loss of intellectual capital and employment skills, which would be needed to rebuild.[25] In this light, plague-like environmental events might be understood as embedded with contexts of human systems failures and conflicts.[26] Looking forward, one must question the limits of protection as the natural disasters that insurers call "acts of God" become more common and more extreme. Unsurprisingly, the insurance industry, led by some of the largest re-insurers who assume the financial risks of catastrophic events, is using its own accounting models to assess potential consequences.[27]

The redistribution of rainfall and the increased flooding that some will face in the future provide a way to look more closely at vulnerabilities to plague-like events. As a hypothetical example, say a storm that was once expected to hit an area every 50 years comes every 30 years. At a minimum, this condition may prompt a reevaluation of risk and a commensurate rise in the cost of premiums. It is conceivable that if the patterns change rapidly, insurance companies (and re-insurance companies, which insure the insurers) will not be able to adequately adjust their premiums and may, therefore, be financially overexposed. Not simply "bad for business" in an abstract way, this situation could result in delayed recovery payments to the insured or, in a worst-case situation, insurance companies defaulting on policies. Adapting to the risks associated with shifting rainfall patterns becomes more complex when it is coupled with increasing land development and urbanization. During a rainfall event, some water is absorbed into the ground while the rest runs off into streams and rivers. The amount of water absorbed is a function of the underlying soils, and when the soil has reached full saturation, all of the remaining water runs off. The speed of runoff is a function of surface texture and slope. (Water will move faster on steep and slick surfaces than it will on relatively flat, "lumpy" surfaces.) Impervious surfaces, which include not only exposed bedrock but also rooftops and concrete and asphalt roads, allow no rainwater infiltration into the ground. Moreover, they shed the water very fast. This means that the more impervious the surfaces within a watershed, the higher streams or rivers will flow, and the faster the peak flow will occur. In a context of unknown event probabilities and increasing costs of repair, insurers may be less willing to offer policies. It is also possible that in the future the costs of replacement will be too much for an insurer, government, or society to bear.

The response to the limits of current social systems—including insurance—may be to reprioritize intelligence activities and to restructure our notion of civil defense. To draw on an example from mainstream media, the Discovery Channel, an American cable television station, produced a television program called "Super Volcano," about the consequences of an eruption of the caldera under Yellowstone National Park, had the United States Geologic Survey becoming a much more prominent organization. The response may also include rethinking how we build in areas that are vulnerable to destructive events. For example, as a result of the recent tsunami that hit Southeast Asia, a new kind of housing is being built.[28] Although the tsunami was not a result of climate change, this kind of approach could also be appropriate for areas prone to coastal flooding that results from hurricanes.

Plants (and Plans)

Included in the category of plants, which begin with a seed that contains an image of the future, are human plans, which similarly begin from a preestablished blueprint or set of expectations. Although it is commonly said that there is no use in complaining about the weather, because nothing can be done about it, people do make plans for climate. Again, we can distinguish between plans to influence the process of climate change and plans to adapt following climate change.

When we currently discuss plans to influence climate change, we usually talk of those to limit greenhouse gas emissions and thereby reduce the likelihood of human-induced climate variability. The Kyoto Protocol is such a plan. Negotiated in December 1997 in Kyoto, Japan (hence the name of the document), and brought into effect in February 2005, its goal was the "stabilization of greenhouse gas concentrations in the atmosphere at a level that would prevent dangerous anthropogenic interference with the climate system."[29] The target is an overall 5.2 percent reduction (as an aggregate for industrial nations) in the emission of CO_2, methane, and other greenhouse gases relative to levels in 1990.[30] However, it should be noted that not each nation is required to contribute, nor even to reduce its output. For example, Australia—which has yet to ratify the Protocol—negotiated a rise in its emissions. Furthermore, China (the second-largest emitter of greenhouse gases), India, and other developing economies are not obliged to reduce emissions. The United States, the largest emitter of greenhouse gases, has signed the Protocol but not ratified it. As expressed in the unanimous (95–0) consent of the Senate in the Byrd-Hagel Resolution, there are concerns that the Protocol would harm the U.S. economy and that it does not include binding timetables for developing nations with expanding economies to meet target emissions. The rationale of the former concern speaks to the difficulty of separating long-term environmental sustainability with short-term economic prosperity. And as such, it relates to the difficulties of pursuing strategies that allow for a diversity of future options, which Stephen Schneider highlighted. The latter concern for not justifying the Protocol might be explained by the more base political worry of "freeloading," of allowing some to benefit while others pay the costs. Without U.S. participation and with the exemption of China, it seems difficult to image that the Kyoto Protocol will achieve its ends; however, that said, individual states within the United States are making their own plans to reduce greenhouse gas emissions.[31] Many other efforts toward minimizing climate change or climate stabilization are those which could be called "no regrets." As discussed in chapter 1, these may be the best alternatives; however, like the Kyoto Protocol, they each present long-term versus short-term trade-offs.

In the context of a discussion of security, it should also be noted that late twentieth and early twenty-first century climate stabilization efforts have not been the only actions proposed. In earlier eras, people spoke of plans to change the climate in order to relieve inhospitable conditions, expand water supplies, and improve food production. At a relatively small scale of physical intervention, farmers have used the locations of hedgerows, woodlots, and even large barns to alter microclimate and thereby increase the yield and profitability of some crops.[32] It has been recognized since the days of ancient Greece and Rome that clearing land of tree cover tends to warm the earth and that the many efforts of many farmers can collectively change the climate of regions.[33] Closer to our own era, writers in the nineteenth century noted that the widespread efforts to settle land in the expanding United States was altering regional climate within a relatively short span of time.[34] The impacts of these changes received mixed reviews. Some saw them as a clear sign of progress, whereas others saw a warning for recklessly tampering with nature. Regardless of whether one was for or against the new weather, it was not idle conversation. Of course, similar changes would have taken place in parts of the world that had long been cleared and planted, and these were also probably noticed by those who worked the land. The transformation in the United States, though, was perhaps more readily recognized and commented upon given the (mistaken) European impression of the continent as "virgin land." The consequences of such incremental and cumulative changes continue today and are being increasingly incorporated in the most advanced climate models.[35]

Distinguishable from these purposeful but individually small actions (*plans* with a lowercase *p*) that can cumulatively affect the climate are those Plans (with a capital P) that involve large scale alterations of landscape. The scale of these plans—Plans—raise the level of action to that of governments and in doing so lead to questions of political power. In some cases, changing the climate was seen not only as a way to organize or to optimize conditions for growth; it was also perceived as a way to expand control over peoples and territory. In colonial America, physician Hugh Williams envisioned a continent laid out to engineer the climate with "vast tracts of cleared land, intersected here and there by great ridges of uncultivated mountains," which would reduce disease and allow new plants to be cultivated.[36] Even more ambitious was the proposal by French engineers in 1912 to turn the middle of the Sahara Desert into a sea by constructing a 50-mile-long canal from the Mediterranean.[37] The new body of water, which would have been half as large as its northern source, would bring local fishing to populations short on arable land and provide a means of commercial traffic between central Africa and Europe. There were, however, more than just aspirations for increased food supplies and free trade associated with this transformation. As noted in the journal *Scientific American,* "[A] great new colony would be added to the possessions of France, of which the political and economic importance can hardly be overestimated."[38] Central to this move of expansion was making local conditions more appealing to the colonizers: "[T]he most remarkable result of all would be the alteration of the climate of all northern Africa from equatorial extremes of heat to the pleasing temperature of Natal, thus enhancing its value as a place of colonization for Europeans."[39] Separate from climate change imperialism, but nonetheless a topic that bridges the environment and security, were the efforts started in the 1950s to weaponize the weather by developing cloud seeding techniques that would cause floods or snowstorms over enemy locations.[40]

Given the political difficulties of limiting global greenhouse gas emissions, some are making plans to respond to climate change. In structuring an understanding of the kinds of plans that are possible, we must consider adaptive capacity, the ability of a society to respond to change and, in particular, to lower its vulnerability to threats. Nick Brooks and his colleagues make a distinction between two kinds of vulnerabilities: First are generic or general vulnerabilities such as poverty, health status, economic inequality, and elements of governance; and second are contextual vulnerabilities, which are specific to the kinds of hazards that are present in different geographies.[41] For example, the contextual vulnerabilities of a small and isolated rural village in semi-arid Africa to the hazard of drought are different than the contextual vulnerabilities of a coastal city in northern Europe to the hazard of flooding. Reducing vulnerability is a matter of understanding the relationships between generic and contextual factors and between biophysical systems and social constructions. In the case of the African village, the construction of a transportation system built by individual states and coordinated among neighboring states (and perhaps involving international nongovernmental organizations) that would allow faster and more reliable connections to food resources would lower vulnerability. In the case of the European city, reducing vulnerability in The Netherlands might include the state-sponsored improvement of the dikes and levees that currently hold back the sea. (And, indeed, such plans are under way.) In other northern European cities, vulnerability might be reduced by the creation of better building codes and their enforcement on private land owners.

The lower limits of adaptive capacity to respond to climate change are perhaps most evident in two examples from opposite sides of the Pacific Ocean. In the south, the government of Tuvalu has made arrangements to evacuate its citizens to New Zealand should the island nation succumb to rising sea levels. Although the plan would laudably protect human life in times of crisis, it embodies a sense of territorial resignation that is counter to every nation's right of self-determination. As such, the implications of this plan call into question the conventional understanding of identity, state sovereignty, and the norms of politics.[42] A similar existential dilemma is faced in the north by the Inuit, who have petitioned the Inter-American Commission on Human Rights for a filing in support of their position that their Arctic villages are becoming uninhabitable due to rising temperatures. Although this commission has no legal authority, a finding in support of the Inuit may allow them to sue the U.S. government or U.S.-based corporations.[43] The Inuit case raises other issues for international relations and law. Assuming we do not call contributions to atmospheric greenhouse gas concentrations acts of aggression because they are not intended to threaten a person, region, or nation-state—although, admittedly, given the mounting evidence linking global warming to climate change, some would say such an argument is sophistry—can individual liability be assigned? Do those nations that produce the most greenhouse gases have a responsibility to provide for any kind of restitution if negative impacts become manifest? Or is the international community as a whole responsible to reduce conditions to which some nations are vulnerable?

It must also be recognized that efforts to enable or to foster some goals can increase vulnerability in other areas. Such occurrences can be called "constructed contingencies" or what Anthony Giddens[44] and subsequently Ulrich Beck[45] have called, "manufactured uncertainties." When recognized, these vulnerabilities make one aware that not every solution to a problem is a *complete* solution. An example of such a situation—of a boomerang effect—is captured by Karen O'Neill's history of U.S. flood control policy along the Mississippi and Sacramento Rivers.[46] Throughout the nineteenth and early twentieth century, agricultural and commercial interests petitioned the federal government to create and then expand flood control measures, which would open new lands for cultivation and development. Although the government in Washington was initially reluctant to act, it eventually created one of the world's largest and most expensive flood control programs. But more than just a network of earthworks and concrete, the system is a complex cultural composition. The federal government owns the rivers, and the states own the banks; and the process of management has the U.S. Army Corps of Engineers working with local levee districts and contractors. (One is left to wonder who owns the floodwaters.) As a result of this configuration, the program can be seen as substantively contributing to the social construction of the United States by helping to define the federal relationship between the nation's capital and the individual states. One manifestation of this joint responsibility is that the federal government does not practice comprehensive natural resource planning for the riverways. Instead, it concentrates on fulfilling the more narrowly defined mission of maintaining navigable passages while addressing district-level priorities. More acutely, there is a cycle of increasing expansion to the system in which flood protection invites more development, which in turn requires more flood protection. Although the new development is arguably "safe," it incrementally diminishes the capacity of natural flood control means. That is, development in new levee-protected areas that were once too wet or too prone to flooding reduces the natural capacity to absorb and distribute floodwater. And therefore, if or when a disaster does strike an area, the damage is all the greater. As this example makes clear, when making plans (our own "plants of the future"), we should bear in mind that we reap what we sow.

Toward that goal of an expanded understanding of security and also toward thinking more precisely about plans that center on potential responses to environmental change, we suggest a framework of strategies created by the medial ecologist William Haddon Jr.[47] Haddon's interest was in generalizing the principles of preventative medicine so that they might be applied to nonliving as well as living hazards (such as viruses). The defining issue is on the release of energy (broadly defined). It can be checked—to say "controlled" would fall victim to hubris—at 10 distinct moments.

1. Prevent the marshaling of the energy in the first place. One of the examples Haddon gives for this category fits well within the narrow definition of threat-centered security concerns: Prevent the concentration of Uranium-235 for weapons production. More in line with vulnerability-centered human security, he mentions not putting babies in high chairs from where they might fall to the floor. In thinking about environmental security, we are not sufficiently advanced to arrest continental plates from colliding or to dissipate the subterranean heat and pressure that leads to volcanic eruptions. However, we might consider the unintentional reduction of solar energy through air pollution—known as "global dimming"—as a move toward this sort of preventative measure. (It should be noted that, thanks to stricter air quality regulations, the skies are clearing in some areas and, as a result, more solar energy is reaching the earth.)[48]

2. Reduce the amount of energy that is marshaled. There have been some intentional actions taken in the past to ameliorate or to decrease the buildup of natural forces and thereby lessen their impacts. A large-scale example is the planting of tree rows

across the American prairies in the 1930s to slow the wind and minimize its contribution to soil erosion.

3. Prevent the release of the energy. A nonenvironmental example of this strategy is locking the safety switch on a loaded gun; although primed for an explosion, the trigger is arrested. Holding back a storm or volcanic explosion is a more complicated matter. However, as an environmental example, we might consider the capture of CO_2 from smokestacks and its sequestering underground. By limiting the amount of greenhouse gases, the atmosphere's capacity to hold heat is reduced.

4. Modify the rate or spatial distribution of the energy release. One of the reasons for the construction of dams throughout the nineteenth and twentieth centuries was the ability to hold back flows in time of flooding. In nature, river deltas distribute and slow the flow of rivers into the sea.

5. Separate, in time or in space, the release of the energy from the vulnerable entity. Haddon cites the physical separation of people and cars by sidewalks and roads and the temporal mediation of flow by traffic lights. Given a warning system, a population can be evacuated from a location threatened by a natural disaster.

6. Separate the release of energy and vulnerable entity by a barrier. This strategy might include the maintenance of natural barrier islands or the construction of artificial islands to absorb the impact of wave action generated by hurricanes.

7. Modify the surface of contact. Haddon, who once worked as Director of the National Highway Safety Bureau, gives an example of the softening and rounding of corners in car design as an example. In terms of larger-scale environmental design, the replacement of impervious concrete and blacktop with porous surfaces and the creation of new wetlands would absorb the shocks of hurricane and severe rainfall events.

8. Strengthen the entity that is vulnerable. As given in the discussion of adaptive capacity for a European coastal city, buildings can be reinforced to withstand greater forces. It is also worth considering how *strength* is to be defined. As has been often asked, which is stronger, a rigid oak tree or a bending willow? The answer depends on the exact nature of the forces applied.

9. Move rapidly to detect, evaluate, and remediate damage after it has occurred to stop it from spreading. Haddon gives the example of a sprinkler that goes off after a fire has started but nonetheless arrests its advance. An environmental analogue might be seen in the identification in the 1970s of the harm caused by fluorocarbons (chlorofluorocarbons, or CFCs, and Hydro-chlorofluorocarbons, or HFCs) to the ozone layer of the atmosphere, which protects us from harmful ultraviolet radiation. In 1985 the Vienna Convention for the Protection of the Ozone Layer was signed. Then in 1987, after scientific proof of the problem was firmly established, the Montreal Protocol brought about the phasing out of the chemicals.[49] This strategy, much more than the others, deals with the social structures of society and adaptive capacity.

10. Repair the damaged entity. In most cases, an impaired object is more susceptible to (further) damage or degradation than a unimpaired object. The restoration may involve a reengineering to include the strategies of strengthening or modifying its form. With regard to reducing environmental vulnerability, native vegetation that has been stripped away by a catastrophic flood might be intentionally replanted immediately rather than waiting for natural regrowth.

Haddon qualifies his list by noting that there is no obvious reason why the order in which he lists the strategies should dictate which strategies should be implemented first; however, he adds, "Generally speaking, the larger the amounts of energy involved in relation to the resistance to damage of the structures at risk, the earlier in the countermeasure sequence must the strategy lie."[50] Thus, in the case of the extreme powers of nature, we would do best by not adding to the buildup of energy.

Plays

The uncertainties of plays, of the drama of life, stem from the ability to exercise free will. The topic of how and why people take action is a profoundly complex task and has long been studied by social scientists and poets. Although articles and books on global warming discuss how people have influenced the climate, there has also been the premise that climate influences the actions of people and the organization of societies. In the earliest surviving text of Greek science, Hippocrates argued that climate was a pervasive factor on the development of human character.[51] And it has since been argued that the climate was the single most important topic in geopolitical history up to the Industrial Revolution.[52] Today such sweeping generalizations are rarely made, although linkages between the influences of climate on human action are still asserted. For example, Emily Oster has found a correlation in the cooling of the Little Ice Age and the scapegoating of women as witches in Europe during the sixteenth and seventeenth centuries. She argues that the economic downturn and food shortages provided a macro scale reason for local persecutions.[53] Less encompassing in causal scope, but not less violent in effect, Edward Miguel has cited income shocks related to crop failures, resulting from extreme rainfall events, as a reason for killing elderly women as witches in present-day Tanzania.[54] More commonly, there are efforts to understand how changes in climate have affected individual settlements or civilizations. Among the most recent offerings is Jared Diamond's *Collapse: How Societies Choose to Fail or Succeed,* which looks at choices—typically poor choices—made by cultures when faced with environmental change.[55] It should be emphasized that although some theses give greater weight to the influence of the environment on human action, none go so far as to argue explicitly that the environment is the final or ultimate cause for a given act. That is, no member of the jury has accepted the excuse "Gaia made me do it."

Our interest in plays of the future is more narrowly how people may respond to security concerns in the context of environmental change. Scholars in fields ranging from history[56] to literary criticism[57] have suggested ways to understanding courses of action, and any of these might be employed to clarify assumptions. For Boulding, there are three basic categories of social interaction that drive human drama: threats (do as coerced under the threat of force), exchange (do as proposed for mutual benefit or gain), and love, along with its opposite, hate (do as motivated based on a sense of identity).[58] When we apply this categorization of action (or one of the other categorizations) to stories of environmental change, we are faced with a definitional problem. Given that climate change can affect so many sectors of society and that the establishment of security can come in so many ways, who are our actors?

As a starting point, we suggest thinking in terms of nations (which are still the conventional unit of analysis for security studies and international relations), but thinking of nations as being comprised of several groups. Such an approach to scenario creation was developed by William M. Jones at the RAND Corporation. His particular goal was to structure thought on a class of military involvements in areas and situations in which the regional powers—the apparent nation-state stakeholders—did not have clearly stated or defined interests.[59] In these cases, it was necessary to understand how other powers might come to define their own interests and then respond to the immediate problem. The understanding of a position on an issue thus becomes a prerequisite for speculating on the plans that might be taken to address it. The method proposes thinking about governments not as monolithic constructions, but as a composition of institutions or interest blocks, each of which has its own operational means and desired ends. As such, the nature of a given nation (its structure and the strength of individual elements) and the nature of the situation at hand both come into consideration. This two-part understanding of action with (or among) other nations can be seen to parallel by considering the two-part understanding of efforts to reduce vulnerability—physical context and social capability (discussed previously). This model, which allows one nation to speculate on the thinking of another, might also be used to consider its own options. For example, in the Schwartz-Randall scenario (chapter 2 and Appendix One), the United States becomes "fortress America" and, in order to protect its own natural resources, it stops river flow to Mexico. Simulating or gaming different constituencies within the country could help to show if or how such a decision is reached.

Jones lists five such institutions that are common to all modern nation-states and suggests what can be considered behavioral tendencies:

1. The foreign ministry. It is conservative when considering change given the needs of its regular contact with counterparts from other nations.
2. The internal administrative bureaucracy. It will shift its position on international problems depending on domestic needs.
3. The military structure. It has a tendency toward "professionalism" and a desire for freedom from political influence. It seeks increased technical and numeric capabilities in the face of enemy threats.
4. The technical and industrial managerial group. Loosely structured, it desires access to domestic and foreign natural resources and markets. It is impatient with ideology

and selfish institutional goals. For this actor, we suspect that Jones might have been thinking of the influence of the "military-industrial complex" about which President Dwight D. Eisenhower warned the nation. (And as a RAND employee, he could rightfully be considered a member of that complex.) Today, the process of globalization has expanded and accelerated the capabilities of non-state actors.[60]

5. The political control structure. It desires national command and views other national institutions as technical service agencies whose recommendations may (or will) inadequately consider ideological and political factors.

In applying this method, it should be recognized that although each kind of institution may have a behavioral tendency, defining the position of an individual institution is not obvious. For example, the pressures within the military structure to train intensively in order to maintain combat readiness can be in conflict with the need to conserve facilities that allow training to take place. Hence, how it might position itself on a given issue relating to environmental change will depend on a case-by-case assessment of short-term versus long-term needs. Moreover, it should be understood that over time the relative influence of each of these institutions will change, waxing and waning based on the general concerns of a society, the details of an individual problem, and the effectiveness of the leadership within each of the five groups.

COMPOSITION AND COMPREHENSION

In presenting an understanding of global warming and climate dynamics in terms of security, we have purposefully tried to eschew the argument that potential environmental change should be met with a militarized response. The problems are too broadly distributed and the consequences are too deeply penetrating for such an approach to be successful. Still, almost any discussion of security leads back to issues of mortal struggle, which are often framed in terms of armed conflict. Despite this distillation, or perhaps because of it, such framings sometimes reveal more general concerns. Literary critic Kenneth Burke observed that "when writing of an anticipated war, the artist must select his material out of the past and the present. All anticipation is such selection, whether it involves one's forebodings about an international calamity or one's attempt to decide whether the red sky at night will be shepherd's delight and the red sky at morning will be shepherd's warning."[61] Although Burke was focused on the representation of war, we might extend his argument to include the representation of other anticipated conditions that can affect entire generations. Our expectations of global climate change likewise involve a sense of foreboding and an attempt to understand the significance of what can be weak signals from nature. If we are to prevent or adapt to climate change, then we must create representations that allow us to come to terms with it.

Planets, plagues, plants (and plans), and plays are each a different sort of anticipation. Each carries specific kinds of uncertainty, and therefore each demands its own set of methods for investigation. For planet-like systems, the sciences tell us what we know, and more importantly, what we can expect. There is a kind of assuredness that comes with such knowledge, but unfortunately, few things behave like planets. The sciences also tell us much about the plagues, which can devastate families, societies, and civilizations. However, here we must also recognize that our own actions can influence the location, timing, frequency, intensity, and duration of these occurrences. Furthermore, the relationships between what we do and how nature will respond are not as well understood as we would like. Our plans are, in part, based on what we know about planets and plagues. They are also based, in part, on our social priorities and available means.

We see three benefits in applying Haddon's framework of energy flows to explore plans of the future. First, as a way to structure potential actions, Haddon's framework provides a comprehensive and systematic approach to identifying options that will reduce harm. Second, this framework recognizes that there are different means to achieve the same end. Haddon's conception addresses problems as occurring over time and recognition that there are multiple opportunities to intervene before, during, and after a given event. Importantly, Haddon accepts that action at each and every step is neither always feasible nor even possible. Moreover, intervening at a later stage will not necessarily achieve the same results as intervening earlier. Third, the outlook of public health helps to move the discussion of security away from a militarized mind-set. At a minimum, it is a productive opening metaphor; but, more substantially, it might be a model of operations.

Similarly, although Jones's approach to scenario creation was developed during the Cold War in anticipation of potential hot combat, it has two qualities that are useful for considering the plays embedded in climate change futures. First, it is predicated on the idea that whatever situation is examined, there are no preestablished interests. We do not know when or how environmental problems will become crises, nor do we know how or when others will respond. Second, though predicated on security concerns, the approach openly and explicitly includes nonmilitary and even nongovernmental entities.

Finally, we should keep in mind some other words by E. M. Forster: "If we would grasp the plot we must add intelligence and memory."[62]

NOTES

1. Pierre Wack is often cited as heralding scenario use through his two seminal articles: "Scenarios: Uncharted Waters Ahead," *Harvard Business Review* 63 (September–October 1985): pp. 72–89, and "Scenarios: Shooting the Rapids," *Harvard Business Review* 63 (November–December 1985): pp. 139–50. For an overview of the development of scenario use in the corporate sector, see Art Kleiner, *The Age of Heretics: Heroes, Outlaws, and the Forerunners of Corporate Change* (New York: Currency Doubleday, 1996).

2. For background on the study and practice of futures analysis, see Wendell Bell, *Foundations of Futures Studies, Volume 1: Human Science for a New Era* (New Brunswick, NJ: Transaction Publishers, 1997), and Wendell Bell, *Foundations of Futures Studies, Volume II: Values, Objectivity, and the Good Society* (New Brunswick, NJ: Transaction Publishers, 1997).

3. Clayton Roberts discusses three kinds of logic historians use to understand purposeful action: the logic of the situation (How does a given context suggest or demand certain actions?); the logic of dispositional traits (How does an individual typically behave?); and the logic of subsequent actions (How does what happened later in the course of events help to reveal what came before it?). When constructing scenarios, all is fiction, so the logic of subsequent actions would be replaced by a logic of anticipation. For Roberts's framework, see Clayton Roberts, *The Logic of Historical Explanation* (University Park: Pennsylvania State University Press, 1996), pp. 160–98.

4. Peter Schwartz and Doug Randall, *An Abrupt Climate Change Scenario and Its Implications for United States National Security* (Emeryville, CA: Global Business Network, October 2003), p. 16.

5. E. M. Forster, *Aspects of the Novel* (London: Harcourt, Inc. [1927, 1955] 1985), p. 87.

6. Bertrand de Jouvenel, *The Art of Conjecture,* trans. Nikita Lary (New York: Basic Books, 1967), pp. 17–18.

7. See Wendell Bell, *Foundations of Futures Studies, Volume 1.*

8. Kenneth E. Boulding, "World Society: The Range of Possible Futures" in *The Future: Images and Processes,* ed. Elise Boulding and Kenneth E. Boulding (Thousand Oaks, CA: Sage Publications, 1995), pp. 39–56; originally published in Donald J. Ortner (ed.), *How Humans Adapt: A Biocultural Odyssey* (Washington, DC: Smithsonian Press, 1983).

9. For background on the influence of the Earth's orbit on climate change, see John Imbrie and Katherine Palmer Imbrie, *Ice Ages: Solving the Mystery* (Short Hills, NJ: Enslow Publishers, 1979); Spencer R. Weart, *The Discovery of Global Warming* (Cambridge, MA: Harvard University Press, 2003).

10. Robert J. Lempert, Steven W. Popper, and Steven C. Banks, *Shaping the Next 100 Years: New Methods for Quantitative Long-Term Policy Analysis,* RAND Report RM-1626 (Santa Monica, CA: RAND Corporation, 2003); Herman Kahn, William Brower, and Leon Martel, *The Next 200 Years: A Scenario for America and the World* (New York: William Morrow and Company, 1976).

11. Imbrie and Imbrie, p. 178.

12. William F. Ruddiman, *Plows, Plagues, and Petroleum: How Humans Took Control of Climate* (Princeton, NJ: Princeton University Press, 2005).

13. Ibid., p. 171.

14. Arthur C. Clarke, "Beyond 2001," *AsiaWeek*—Millennium Special Edition (August 20/27, 1999), www.pathfinder.com/asiaweek/99/0820/cs2.html. Reprinted in *Readers Digest* (February 2001).

15. U.S. National Air and Space Administration (NASA), "The 1991 Mt. Pinatubo Eruption Provides a Natural Test for the Influence of Arctic Circulation on Climate" (March 12, 2003), www.nasa.gov/centers/goddard/news/topstory/2003/0306aopin.html.

16. Henry M. Stommel and Elizabeth Stommel, *Volcano Weather: The Story of 1816, the Year without a Summer* (Newport, RI: Seven Seas Press, 1983); C. R. Harington, ed.,

The Year without a Summer? World Climate in 1816—Papers from the "Workshop on World Climate in 1816" (Ottawa: Canadian Museum of Nature, 1992).

17. Jen Bissell, "A Comet's Tale: On the Science of Apocalypse," *Harper's* (February 2003): 35.

18. Z. W. Kundzewicz and H. J. Schellnhuber, "Floods in the IPCC TAR Perspective," *Natural Hazards* 31 (January 2004): pp. 111–28; Robert J. Nicholls and Jason A. Lowe, "Benefits of Mitigation of Climate Change for Coastal Areas," *Global Environmental Change—Human and Policy Dimensions* 14 (October 2004): pp. 229–44.

19. Ning Zeng, "Drought in the Sahel," *Science* 302 (November 7, 2003): pp. 999–1000; Kate Ravilious, "Ice Melt May Dry Out U.S. West Coast," *New Scientist* 182 (April 10, 2004): p. 17.

20. Gerald A. Meehl and Claudia Tebaldi, "More Intense, More Frequent, and Longer Lasting Heat Waves in the 21st Century," *Science* 305 (August 13, 2004): pp. 994–97.

21. James Randerson, "Climate Change Blamed for Upsurge in Disease," *New Scientist* 182 (June 19, 2004): pp. 8–9.

22. M. Totten, S. I. Pandya, and T. Jason-Smith, "Biodiversity, Climate, and the Kyoto Protocol: Risks and Opportunities," *Frontiers in Ecology and the Environment* 1 (June 2003): pp. 262–70.

23. Peter L. Bernstein, *Against the Gods: The Remarkable Story of Risk* (New York: John Wiley & Sons, 1996).

24. Walter G. Green III, "The Future of Disasters: Interesting Trends for Interesting Times," *Futures Research Quarterly* 20 (Fall 2004): pp. 59–68, this note p. 63.

25. Tom Barbash, *On Top of the World: Cantor Fitzgerald, Howard Lutnick, & 9/11: A Story of Loss & Renewal* (New York: HarperCollins, 2003).

26. Walter G. Green III and Suzanne R. McGinnis, "Thought on the Higher Order Taxonomy of Disasters," *Notes on Extreme Situations,* (September 2002), p. 1–5.

27. Mojdeh Keykhah, "Global Hazards and Catastrophic Risk: Assessments in the Reinsurance Industry," in *Assessments of Regional and Global Environmental Risks: Designing Processes for the Effective Use of Science in Decisionmaking,* ed. Alexander E. Farrell and Jill Jager (Washington, DC: Resources for the Future, 2006), pp. 242–58; Michael Tucker, "Climate Change and the Insurance Industry: The Cost of Increased Risk and the Impetus for Action," *Ecological Economics* 22 (August 1997), pp. 85–96.

28. Ken Gewertz, "GSD Students Win Tsunami Design Award: Devise Strategy for Development, Permanent Housing," *Harvard University Gazette,* May 26, 2005, www.news.harvard.edu/gazette/2005/05.26/01-srilanka.html.

29. United Nations Framework Convention on Climate Change, Article 2. Full text of the document is available at unfccc.int/essential_background/convention/background/items/2853.php.

30. To view the final draft of the Kyoto Protocol to the United Nations Framework Convention on Climate Change, see http://unfccc.int/resource/docs/cop3/l07a01.pdf.

31. Roger Harrabin and Steve Hounslow, "'Gas Muzzlers' Challenge Bush," BBC News, November 3, 2005, news.bbc.co.uk/2/hi/americas/4400534.stm.

32. John R. Stilgoe, "Skewing Private Climate," in *Landscape and Images,* ed. John R. Stilgoe (Charlottesville: University of Virginia Press, 2005) pp. 170–92; originally published as "A History of Microclimate Modification 1600 to 1980," in *Energy Conserving Site Design,* ed. Gregory McPherson (Washington, DC: American Society of Landscape Architects, 1984), pp. 2–16.

33. Clarence J. Glacken, *Traces on the Rhodian Shore: Nature and Culture in Western Thought from Ancient Times to the End of the Eighteenth Century* (Berkeley: University of California Press, 1967), pp. 129–30.

34. Ibid., p. 689ff.

35. Roger A. Pielke Sr., "Land Use and Climate Change," *Science* 310 (December 9, 2005), pp. 1625–26; Johannes J. Feddema, Keith W. Oleson, Gordon B. Bonan, Linda O. Mearns, Lawrence E. Buja, Gerald A. Meehl, and Warren M. Washington, "The Importance of Land-Cover Change in Simulating Future Climates" *Science* 310 (December 9, 2005), pp. 1674–78.

36. Hugh Williamson, "An Attempt to Account for the Change of Climate, which Has Been Observed in the Middle Colonies in North-America," *Transactions of the American Philosophical Society,* Volume 1, 2nd ed. (1789), pp. 336–435; this quote p. 343, cited and discussed in Glacken, p. 660ff.

37. G. A. Thompson, "A Plan for Converting the Sahara Desert into a Sea," *Scientific American* 107 (April 10, 1912), pp. 114, 124–25.

38. Ibid., p. 114.

39. Ibid.

40. Spencer R. Weart, *The Discovery of Global Warming* (Cambridge, MA: Harvard University Press), p. 23; Schneider, p. 210ff.

41. Nick Brooks, W. Neil Adger, and Mick Kelly, "The Determinants of Vulnerability and Adaptive Capacity at the National Level and the Implications for Adaptation," *Global Environmental Change* 15 (2005): 151–63; Nick Brooks, "Vulnerability, Risk and Adaptation: A Conceptual Framework," Working Paper 38—Tyndall Center for Climate Change Research (November 2003).

42. Jon Barnett and W. Neil Adger, "Climate Dangers in Atoll Countries," *Climatic Change* 61:3 (2003): pp. 321–37.

43. Tim Flannery, *The Weather Makers: How Man Is Changing the Climate and What It Means for Life on Earth* (New York: Atlantic Monthly Press, 2005), p. 286.

44. Anthony Giddens, *The Consequences of Modernity* (Cambridge, UK: Polity, 1990); *Beyond Left and Right* (Cambridge, UK: Polity, 1995).

45. Ulrich Beck, "Risk Society Revisited," in *World Risk Society,* Ulrich Beck (Cambridge, UK: Polity), pp. 133–52.

46. Karen M. O'Neill, *Rivers by Design: State Power and the Origins of U.S. Flood Control* (Durham, NC: Duke University Press, 2006).

47. William Haddon Jr., "On the Escape of Tigers: An Ecological Note," *American Journal of Public Health and the Nation's Health* 60:12 (1970): pp. 2229–34; originally published in *Technology Review* 72:7 (May 1970), pp. 4453.

48. Quirin Schiermeier, "Cleaner Skies Leave Global Warming Forecasts Uncertain," *Nature* 435 (May 12, 2005), p. 135; Martin Wild, Hans Gilgen, Andreas Roesch, Atsumu Ohmura, Charles N. Long, Ellsworth G. Dutton, Bruce Forgan, Ain Kallis, Viivi Russak, and Anatoly Tsvetkov, "From Dimming to Brightening: Decadal Changes in Solar Radiation at Earth's Surface," *Science* 308 (May 6, 2005): pp. 847–50; R. T. Pinker, B. Zhang, and E. G. Dutton, "Do Satellites Detect Trends in Surface Solar Radiation?" *Science* 308 (May 6, 2005): pp. 850–54.

49. Edward A. Parson, "Grounds for Hope: Assessing Technological Options to Manage Ozone Depletion," in *Assessments of Regional and Global Environmental Risks: Designing Processes for the Effective Use of Science in Decisionmaking,* ed. Alexander E. Farrell

and Jill Jager (Washington, DC: Resources for the Future, 2006), pp. 227–41; Detlef F. Sprinz, "Comparing the Global Climate Regime with other Global Environmental Accords," in *International Relations and Global Climate Change,* ed. Urs Luterbacher and Detlef F. Sprinz (Cambridge, MA: MIT Press, 2001), pp. 247–77.

50. Haddon, p. 2232.

51. Glacken.

52. Captain Charles Konigsberg, "Climate and Society: A Review of the Literature," *The Journal of Conflict Resolution* (Special Issue on The Geography of Conflict) 4 (March 1960): pp. 67–82.

53. Emily Oster, "Witchcraft, Weather and Economic Growth in Renaissance Europe," *Journal of Economic Perspectives* 18 (Winter 2004): pp. 215–28.

54. Edward Miguel, "Poverty and Witch Killing," *Review of Economic Studies* 72 (2005): pp. 1153–72.

55. Jared Diamond, *Collapse: How Societies Choose to Fail or Succeed* (New York: Viking, 2005).

56. Roberts.

57. Kenneth Burke, *A Grammar of Motives* (Berkeley: University of California Press, 1969).

58. Kenneth E. Boulding, "The Relations of Economic, Political, and Social Systems," *Social and Economic Studies* 11 (1962): pp. 351–62; Kenneth E. Boulding, *Ecodynamics* (Beverly Hills, CA: Sage Publications, 1978).

59. William M. Jones, "Fractional Debates and National Commitments: The Multidimensional Scenario," RAND Memorandum RM-5259-ISA (Santa Monica, CA: The RAND Corporation, 1967).

60. Richard A. Higgott, Geoffrey R. D. Underhill, and Andreas Bieler, eds., *Non-State Actors and Authority in the Global System* (New York: Routledge, 2000).

61. Kenneth Burke, "War, Response, and Contradiction," in *The Philosophy of Literary Form: Studies in Symbolic Action,* 3rd ed., Kenneth Burke (Berkeley: University of California Press, 1973), pp. 234–57; this note p. 237.

62. Forster, p. 87.

— 7 —

The End of the Anthropocene

Men and nations behave wisely once they have exhausted all the other
alternatives.

Abba Eban

It would be classic understatement to declare, at this work's conclusion, that we
have entered an age of immense change. Yet, ultimately, we have not adequately
assessed—or appreciated—the vicious cycle we have entered. Human impact
has led to climate change, and such anthropogenically induced change will
further affect human ability to act, adapt, or respond. Ironically, in the 1960s
the atmospheric scientist and chemist Sir James Lovelock posited the "Gaia hypothesis," suggesting that the earth itself is a self-regulating superorganism that
seeks balance and sustainable conditions for continuity of life. If the earth falls
out of balance, there will be an attempt to return.

Granted, there are cataclysmic "natural" events—such as the Asian tsunami
of December 2004. Yet, in the aftermath of Hurricane Katrina in 2005—which
devastated the Gulf Coast of the United States and where increased sea and
surface temperatures *may* have led to intensified storm effects—the debate has
slowly, but definitely, emerged as to whether humankind has contributed to
knocking the earth "out of balance."

Some potential outcomes, not necessarily about to happen in this decade—
or even in this century—are nonetheless staggering. In 1927, the Serbian geo-
physicist Milutin Milanković developed his astronomical theory of climatic
variation of the seasons based on the earth's elliptical rotation around the sun

(and the earth's own rotation on a displaced axis).[1] (Recall that the Milanković Cycle, alternatively known as the "Milanković Mechanism," is discussed at greater length in chapter 6.) Yet evidence has also since emerged that increased greenhouse carbon dioxide (CO_2) can lead to "positive feedback" in relation to ice volume. (Moreover, changes in CO_2 seem to affect both southern and northern hemispheres synchronously and in phase with northern ice.)[2] In light of the generally accepted evidence that over the twentieth century, global temperatures have risen by about one degree (Fahrenheit), there is emerging evidence that the depletion of ice volume for the Greenland ice sheet—along with Antarctica, the second massive ice sheet, in the southern hemisphere—is actually accelerating. If true, then enormous cascading effects, compounded by the Milanković Mechanism, could produce complex, even chaotic, and entirely unpredictable outcomes.

Consider, for example, that if the entire Greenland ice sheet vanished, global sea levels could rise as high as 23 feet, engulfing much of Bangladesh and coastal communities on the Eastern seaboard of the United States. Even more frightening, the Antarctic ice sheet, should it disappear, has the potential to raise global sea levels 215 feet.[3] Given such staggering potentials, one can only ask if the earth itself—or, perhaps the earth's namesake, the Greek goddess Gaia—is seeking revenge.

In the first quarter of the twentieth century, the Russian geochemist Vladimir Ivanović Vernadsky, the French Jesuit P. Teilhard de Chardin, and E. Le Roy created the concept of the "noösphere," a world powered by human ingenuity and technology that would better the future and the environment. Theirs was a concept full of optimism. Yet in the first decade of the twenty-first century, the Nobel laureate Paul Crutzen and his colleague Eugene Stoermer suggested a much darker notion of human impact on the environment by arguing that we now live in the age of the "Anthropocene." No longer, Crutzen and Stoermer suggest, should we consider ourselves living the age of the Holocene—spanning almost 12 millennia (in calendar, versus radiocarbon, years) and encompassing the entire course of human civilization. To the contrary, something entirely different has occurred:

To assign a more specific date to the onset of the "anthropocene" seems somewhat arbitrary, but we propose the latter part of the eighteenth century, although we are aware that alternative proposals can be made (some may even want to include the entire holocene). However, we choose this date because, during the past two centuries, the global effects of human activities have become clearly noticeable. This is the period when data retrieved from glacial ice cores show the beginning of a growth in the atmospheric concentrations of several "greenhouse gases," in particular CO_2 and CH_4. Such a starting date also coincides with James Watt's invention of the steam engine in 1784.[4]

An even darker proposition is to consider whether we have entered the "late" stage of the Anthropocene—when by our own actions, we have determined, if not sealed, our own fate. Among some environmentalists (perhaps those who might be most commonly called "extreme"), there has emerged the argument that the actions, if not the arrogance, of humanity have proven the greatest single threat to ecological sustainability. To save the earth, such logic goes, the threat must be eliminated; in other words, as part of the natural evolutionary process, humans will disappear from the planet. The unique distinction, none-theless, would be that humanity's own actions led to humanity's extinction.

One degree global temperature rise in the space of a century . . . . According to the 2001 United Nations Intergovernmental Panel on Climate Change, projected temperature rises could be as great as 10.4° F in the twenty-first century.[5]

Nothing is more powerful than a reality that has reached its time. Yet, it is not as though we have not been warned. Although we may have considered the enormous complexities of the warning signs over the last four decades of the twentieth century, we have seen few positive effects in our attempt to moderate emissions, shift technologies, or reduce energy consumption. And there have been roadblocks along the way.

SCIENCE AND POLITICS MUST MIX IT UP

James Hansen, Director of NASA's Goddard Institute for Space Studies, was one of the few scientists to engage in climate modeling in the 1970s; the results of the Goddard Institute's early work were alarming enough that the administration of then-President Jimmy Carter called on the National Academy of Sciences to investigate further. The Academy undertook its first major study of the effect of adding CO_2 to the atmosphere in 1979. Led by Jule Charney of the Massachusetts Institute of Technology, the Ad Hoc Study Group on Carbon Dioxide and Climate provided some blunt assessments in its conclusion: "If carbon dioxide continues to increase, the study group finds no reason to doubt that climate changes will result and no reason to believe that these changes will be negligible. . . . We may not be given a warning until the CO_2 loading is such that an appreciable climate change is inevitable."[6] The effect of adding CO_2 to the atmosphere, the report suggested, would be to knock the earth out of energy balance necessary for ecological sustainability and continued biodiversity.

In 1988, James Hansen testified before the U.S. Senate that, based on computer models and temperature measurements, he was "99 percent" sure that anthropogenically caused greenhouse effect had been detected and was changing the climate. His statement was widely covered by the media and brought the term "global warming" to the general public's attention for the first time.[7] Notably—perhaps the statement he was most savaged for in later critiques—he stated: "It is time to stop waffling so much and say that the evidence is pretty strong that the greenhouse effect is here."[8] By the 1990s, Hansen, after reviewing data taken on ever-accelerating flow rates in the Jakobshavn Isbrae ice river (part of the Jakobshavn Ablation Region in the Greenland Ice Sheet), argued that if greenhouse gas emissions could not be controlled, the disintegration of the Greenland ice sheet could begin in a matter of decades.[9]

As a scientist and administrator concerned with integrity and discipline, Hansen has been scrupulous in his work to insist on the lack of absolute certainly in climate change modeling and research. In a 1999 essay, for example, he begins with an epigraphic quote that clearly encompasses his own philosophic and scientific approach to climate complexity:

The only way to have real success in science ... is to describe the evidence very carefully without regard to the way you feel it should be. If you have a theory, you must try to explain what's good about it and what's bad about it equally. In science you learn a kind of standard integrity and honesty.[10]

—*Richard Feynman*

If anything, Hansen has been consistent in his posture and behavior. Although not hesitating to take on sometimes unscrupulous skeptics who have modified or manipulated his arguments,[11] he has also expressed concern that both media and environmental advocates be more careful in their reporting and their rhetoric addressing climate change.[12]

Yet, it may be precisely because of his integrity and discipline that he has been the target of such frequent ridicule, or scorn, by impatient, uninformed, or simply unbalanced skeptics. H. Sterling Burnett, for example, writing for the National Center for Policy Analysis, suggests that "James Hansen, whose 1988 pronouncements started the clamor for action to prevent global warming, wrote in the 1998 *PNAS* [Proceedings of the National Academy of Sciences of the United States of America] that 'the forcings that drive long-term climate change are not known with accuracy sufficient to define future climate change.' So much for being sure."[13] Burnett nonetheless seems, perhaps intentionally, to be missing Hansen's point about the complexity of scientific prediction. Hansen's dilemma, of course, returns us to the concept we visited in chapters 3, 5, and 6 regarding Gidden's "manufactured uncertainty": A paradox occurs when increased knowledge fuels the production of ever more increased uncertainty.[14]

The real savaging of James Hansen, nonetheless, came from within his own organization. First reported in the *New York Times* in late January 2006 with the title "Climate Expert Says NASA Tried to Silence Him," Hansen directly accused Bush administration officials of attempting to silence his call for swift reduction in greenhouse emissions of gases linked to global warming.[15] Dean Acosta, deputy administrator for public affairs at NASA, states that there was no effort to silence Hansen; rather, the effort was meant to "coordinate" NASA responses to interview requests, to present a united agency perspective, and to discourage scientists from the appearance of making policy statements—which should be left to policy makers and appointed spokespersons. Hansen, in contrast, disagreed with this assessment, claiming that "communicating with the public seems to be essential because public concern is probably the only thing capable of overcoming the special interests that have obfuscated the topic."[16]

Earlier in his career, Hansen had made the claim that "science and politics don't mix." On this particular point, however, Hansen may have come to the recognition that—in order to effect change—his original hypothesis about science and politics was most definitely wrong.

SKEPTICS AND THEIR INFLUENCE

We do not mean to suggest that there is no place for skepticism in this debate. Without the necessary doubt that must exist in the presence of uncertainty, one is either driven by faith or, in turn, simply proselytizing canonic belief. As we discuss subsequently, however, there have been some recent manipulations of skepticism regarding climate change that have been put forth in order to prevent effective policy action.

Among prominent doubters, the statistician Bjørn Lomberg, author of *The Skeptical Environmentalist,* has argued that the cost of arresting global warming could be many trillions of dollars more than the cost of global warming itself.[17] S. Fred Singer, author of *Hot Talk Cold Science,* largely denies that there is any credible evidence of global warming. Yet Singer—who charges that most climate scientists are "professional environmental zealots" in search of government-funded research to perpetuate their research—based much of his arguments on satellite data that detected little change in temperatures.[18] As discussed in chapter 5, nonetheless, more recent satellite data *have* detected warming. It is also notable that global warming skeptics do not publish, as a rule, in peer-reviewed scientific journals; rather, with exceptions, such skeptical critiques appear in publications funded with industrial support or through conservative foundations—or in business-focused media such as the *Wall Street Journal*.[19] (Articles in peer-reviewed scientific journals, by contrast, almost overwhelmingly support the evidence for global warming.)[20] Equally, those interested in opposing the evidence against climate change have incentive and ample opportunity for research through industrial support as well.

The novelist Michael Crichton is likely the most well-known popular skeptic regarding climate change. And, although his arguments that the "scientific" claims regarding global warming have been obfuscated and distorted may seem extreme to some, he has been effective in enforcing the principle that, in order to convince an audience, one most be simple, clear, direct, and understandable. Primarily, in his challenges to the UN's Intergovernmental Panel on Climate

Change (IPCC) for its tendency to produce work written by specialists and intended to be read by specialists, he is correct in emphasizing that there are better ways to do business.[21] Here is the basic conundrum we considered under Level 1 ("Clear Enough Future") and Level 3 Uncertainty ("Range of Futures") in chapter 5: The requirement that scientific modeling and data allow for sometimes wide variability versus the need for decision makers to have strict limits within which to make their choices. For science, the requirement is to allow for a multitude of possibilities; yet, to be blunt—especially in Washington—the basic conundrum comes down to this: "Tell me in thirty seconds or less why this matters."[22]

Journalist Ross Gelspan has also suggested that there is far more insidious work at hand in the role of forcing skepticism on the general public regarding climate change and human impact. In a book title that lays out his perspective quite clearly (*Boiling Point: How Politicians, Big Oil and Coal, Journalists, and Activists Have Fueled the Climate Crisis—and What We Can Do to Avert Disaster*), Gelspan's opening lines pull no punches: "It is an excruciating experience to watch the planet fall apart piece by piece in the face of persistent and pathological denial."[23] Mark Hertsgaard published a similarly damning exposé titled "While Washington Slept" in the May 2006 issue of *Vanity Fair*. In describing "powerful pockets of resistance"—even as the scientific evidence regarding climate change grew ever more definitive—Hertsgaard draws on the example of Dr. Frederick Seitz (and others) that followed a strategy on the deniability of global warming similar to that used by tobacco companies for decades to delink a connection between smoking and health effects.[24] In an earlier book titled *The Heat Is On*, Ross Gelspan quotes a 1991 strategy memo as part of a larger corporate campaign to "reposition global warming as theory rather than fact."[25]

To be sure, much popular journal reporting on climate change data and projections is either alarmist in assessment or too focused on the elusive goal of "balance"—rather than objectivity—in coverage.[26] Bill McKibben, author of the first major popular work on global warming, *The End of Nature*, noted in the tenth anniversary edition of the book that, despite the overwhelming evidence gathered through scientific research, much of the coverage on climate change by journalists seems "to be coming from some other planet."[27] And, admittedly, given the hyperbolic media responses to the Schwartz-Randall scenario considered in this work—"Pentagon Warning of a New Ice Age,"[28] "A New Ice Age?"[29] "The Pentagon's Weather Nightmare,"[30] "Pentagon Report Plans for Climate Catastrophe,"[31] "Pentagon-sponsored Climate Report Sparks Hullabaloo in Europe but New Ice Age Unlikely, Bay Area Authors of Study Say,"[32] "'The Sky Is Falling!' Say Hollywood and, Yes, the Pentagon,"[33] "Now the Pentagon Tells Bush: Climate Change Will Destroy Us: Secret Report Warns of Rioting and Nuclear War; Britain Will Be 'Siberian' in Less Than 20

Years; Threat to the World Is Greater than Terrorism,"[34] "Flash! Global Warming May Bring Ice Age By The Next Election"[35]—there may be some credibility to McKibben's claim.

Yet, there are notable exceptions: Andrew Revkin of the *New York Times* (and author of one of the extreme headlines regarding the Schwartz-Randall scenario) remained largely consistent in explaining the technical, and significant, features of scientific research on climate change; and Elizabeth Kolbert, a staff writer for the *New Yorker,* provided the overall clearest and most convincing science journalism regarding human impact in a series of three essays titled "Annals of Science: The Climate of Man" between April and May 2005.[36] Kolbert acknowledged that "there is a surprisingly large—you might even say frighteningly large—gap between the scientific community and the lay community's opinions on global warming." Part of this perception discrepancy, she reasoned, is due to a "well-financed disinformation campaign designed to convince people that there is still scientific disagreement about the problem, when ... there really is quite broad agreement."[37]

It is difficult, of course, to know precisely the nuances and the intricacies of policy agenda and decisions. Occasionally, as noted in chapter 5, glaring examples of gross intervention sometimes do break through—as when Philip Cooney altered or removed descriptions of climate research that government scientists and some senior Bush administration officials had already approved. Yet there can be—and are—even larger influences and forces at play. Paula Dobriansky, U.S. Under Secretary of State for Global Affairs and Democracy, within months of being confirmed by the U.S. Senate, met with ExxonMobil lobbyists and the Global Climate Coalition (GCC; an outspoken and confrontational industry group that actively argued against reductions in greenhouse gas emissions).[38] For her meeting with the GCC, one of Dobriansky's talking points noted that "POTUS [President of the United States] rejected Kyoto, in part, based on input from you."[39] Thus, despite Dobriansky's emphasis on "the rhetoric of scientific uncertainty" as a consistent theme regarding climate change and human impact, it would appear she was far more attuned to the pragmatics of political reality.

In larger contexts, though, as awareness begins to break through, a disparate array of voices has emerged supporting action—rather than further research—on climate change and human impact. Among the more well-known of these voices to advocate action is the Reverend Richard Cizik, who has led a biblically inspired environmental movement known as Christian Care; on occasion, he has been known to cite The Apocalypse of John (Revelation), chapter 11, verse 18: God will "destroy those who destroy the earth." In February 2006, Cizik, among 86 evangelical Christian leaders, released a manifesto titled *Climate Change: An Evangelical Call to Action,* invoking movement against global warming and arguing that the poor and the poorest regions of the world would suffer the greatest human impact as the result of global warming and climate change. (Roman Catholic and Jewish groups have also called for action on global warming.) Notably, James Dobson, leader of Focus on the Family, and Charles Colson were among 22 opposing evangelical leaders who reacted harshly against the call to action and expressed continued support for Bush administration policies. Evangelicals, nonetheless, wield significant political weight in the contemporary political scene in the United States; without their support, little effective political action would take place.

Of equal note, the Free Enterprise Action Fund, which owns less than 1 percent of outstanding shares of General Electric (GE), was able in 2006—with the support of the Securities and Exchange Commission—to stop GE from blocking a resolution put forward by the mutual fund that called on GE to stop advocating global warming regulation.[40] Steve Milloy, the Free Enterprise Action Fund's leader, noted that, among other issues, "even if human activity is altering global climate to some extent, such climate change might actually be beneficial—historically, civilization has fared better in warmer climactic conditions as opposed to cooler climactic conditions."[41] Such comments seem jarringly out of place, however, when one considers how slightly increased temperatures have brought devastating, possibly semipermanent drought to contemporary East Africa. Moreover, Milloy's remarks echo the earlier belief of Nobel Laureate Svante Arrhenius, expressed a century before, that increased temperatures would prove beneficial to the overall global human condition. Then again, as we noted earlier in this work, Arrhenius predicted that it would take 3,000 years to double the CO_2 levels in the atmosphere—and, as we know today, his estimate was off by *28 centuries.*

THE ECONOMICS OF "ECOTERRORISM"

The term *ecoterrorism* can take on chimera-like qualities, representing often more the perspective of the user of the term than any precise definition. For conservatives, many environmentalists are ecoterrorists. For some environmentalists, those who do not accept the need for change in energy consumption, pollution patterns—or who are unwilling to engage, mitigate, or even effectively adapt to the vulnerabilities of global environmental shifts—are, again, ecoterrorists. One of the authors (Liotta), by virtue of being a U.S. citizen, was himself accused of being an ecoterrorist during a recent research trip to South Africa. And, after some reflection, Liotta agreed that there may be some virtue to that argument.

When confronted with the term *ecoterrorism,* the contention was that although terrorists will, on occasion, strike with devastating impact, the U.S. ecoterrorist, by virtue of consumption patterns, energy abuse, and environmental footprints that leave a global shadow, actually has a far wider impact than the aggregate of other terrorist activities. The average American, as noted previously, produces the same greenhouse-gas emissions as 4.5 Mexicans, 18 Indians, or 99 Bangladeshis. Americans are overwhelmingly responsible for the highest per capita greenhouse emissions. And American use of petrochemicals—not only for gasoline but for an infinite of petroleum-based products (including the polymer casing for the laptop computer on which these words are being typed)—far exceeds that of any other state on the face of the earth. In 2002, for example, the United States consumed 7.2 billion barrels of oil—almost four times the amount of the other closest consumers, China and Japan. By 2020, roughly, the United States—if consumption patterns and energy conditions remain unchanged—would burn through more than 10 billion barrels annually, while production would be *less* than two billion barrels annually.[42] This is proving to be a dilemma, not only for oil consumption but for climate change and human impact as well:

Oil consumption in the United States has been steadily rising since Jimmy Carter left office, in 1981. If during that time fuel-efficiency standards for cars and light trucks had been raised by just five miles per gallon, we would now be using one and a half million barrels of oil less each day, and if they had been raised by ten miles per gallon we would be using two and a half million barrels of oil less each day. If fuel-efficiency standards were raised to forty miles per gallon—a level that is eminently achievable with current technology—the United States would save sixty billion barrels of oil over the next fifty years. Simply upgrading the standards for replacement tires so that they match those for tires on new cars would avert the need for seven billion barrels, which is roughly the same amount we could hope to get out of the Arctic Refuge.[43]

Both paradox and conundrum are present in this dilemma. The source of American economic strength lies partly in its extraordinary consumption patterns. Removing, or attempting to quickly alter such patterns, could well assist in the process of a kind of economic suicide.

Yet, equally, it is reasonable to ask how long such devastating impacts could, or should, be allowed to last. What, in other words, are the economic costs of ecoterrorism? The answers are not clear. The Congressional Budget Office, in April 2003, produced *The Economics of Climate Change: A Primer* but notably did not provide specifics on what the costs would be to effect change—partly because this is simply too complex an analysis to project. And though some might argue that it is cheaper to *adapt* to climate change rather than to reduce emissions or mitigate impacts, these arguments remain, at best, conceptual polemics that reflect far more the perspective of the individual argument than any singularly convincing reality. William Nordhaus, perhaps the leading economic authority to consider the cost impact of global warming, has reasoned that the social cost of climate change and human impact amounts to $4 trillion.[44] (Of note, Bjørn Lomberg, author of *The Skeptical Environmentalist,* accepted an even higher estimate—$5 trillion.)[45] Yet Richard Posner counters that such arguments fall wide of the mark in several ways. First, such estimates do not account for the possibility of abrupt, rapid climate change, which could impose far greater, more devastating human costs. Second, Posner argues that comparisons should not be measured against the production of goods and services (the gross domestic product), which amounted to almost $13 trillion in 2005 for the United States;[46] rather, comparison should be made against the stock of physical and human capital that produces such goods and services output—which likely exceeds $100 trillion.[47]

There are limits to understanding the real costs of global warming; to obvert (and update) one of the most beloved phrases of former Senator Everett Dirksen, "A trillion here, and a trillion there, and pretty soon you're talking about *real* money." One frustrating truth is that no amount of cost-benefit analysis or scenario projection can accurately portray whether adaptation is a more viable strategy than changing behavior or mitigation in response to climate change and human impact. Yet these sobering thoughts seem useful to consider here:

At a certain point, though, the changes will become so great that adaptation will become extremely difficult; a five-foot rise in sea levels, for example, would put parts of the state of Florida underwater. If you imagine that sort of scenario being played out all around the globe, it gets pretty frightening. And, as one climatologist pointed out to me, while we are more technologically sophisticated than earlier societies, we are also more sophisticated when it comes to destruction.[48]

FINDING REAL SOLUTIONS

The necessary steps to address the problems of climate change and human impact may well prove the ones most difficult, if not impossible, to take. Though ironic, and persistently true, that we almost *always* wait until the crisis in upon us before we act, the issue here is of such critical importance that waiting any longer will only sustain damage effects in both the short and long term. Primarily, we must generate awareness on the part of the actors that matter in the international system, mobilize this awareness, and marshal actions in ways that make the best of some potentially bad—sometimes fatal—outcomes. There are no guarantees, of course, that with such concentration of effort, solutions will be found in time or will work. All of what is called for here, in light of the failures of environmentalists and politicians to reach common ground, or common advocacy on issues of climate change and human impact, seems admittedly difficult to achieve.

First, there should be recognition that truly affecting issues involving climate change is not a single advocacy focus. Just as climate change can affect a myriad of human issues—from drought to shifting weather patterns to urban living conditions—so, too, should come acknowledgment that climate change will not be drastically affected with the passage of a single statute (e.g., raising automobile fuel standards) or with the ratification of a single international protocol or regime (e.g., Kyoto or its successor). Moreover, given the serial losses on any number of environmental efforts to change consumption and emission, from Kyoto to CAFÉ (Corporate Average Fuel Economy) legislation to even the McCain-Lieberman Act (which only marginally reduces carbon emissions, and is nowhere close to the 50–70 percent required reduction to have major impact), these largely ineffectual measures individually *do* help build toward a sensibility for change, and behavior that reinforces the impression that, over time, unified advocacies and mechanisms can affect carbon reduction.

Second, just as science and politics must mix to effect change, environmental initiatives must be put forward that recognize political realities. While Ross Gelspan, among others, accuses environmental leaders of being too timid to raise alarms about climate change, initiatives that are tabled must be not only technically feasible—they must also be politically possible. As Shellenberger and Nordhaus note,

> What's frustrating about *Boiling Point* and so many other visionary environmental books—from *Natural Capitalism* by Paul Hawken, and Amory and Hunter Lovins to *Plan B* by Lester Brown to *The End of Oil* by Paul Roberts—is the way the authors advocate technical policy solutions as though politics didn't matter. Who cares if a carbon tax or a sky trust or a cap-and-trade system is the most simple and elegant policy mechanism to increase demand for clean energy sources if it's a political loser?[49]

The two authors further note that those advocating technical policy solutions often myopically focus on incremental fixes; almost every environmental leader they interviewed was focused on short-term policy work rather than on long-term strategies. Long-term strategies, nonetheless, help not only establish vision and values, but also set the context for transformative behavior and actions—all of which is crucially necessary to affect climate change and human impact. Political achievability, nonetheless, must be considered a resource in order to secure both the vision of environmental change—and its reality. Conversely, a vision without resources is a hallucination.

Third, the popular Massachusetts politician "Tip" O'Neill is remembered for his saying, "All politics is local"; by extension, all political action will never take root unless local communities accept the incentive for change. Change, in this case, is truly driven from the bottom up. In many cases, where nation-states and international efforts seem unable to agree on what best measures (if any) to take, other forces are at play. In Montana, for example, the (not uncontroversial) governor, Brian Schweitzer, has built a political platform on energy "in-dependency" and finding, quickly, solutions to the current reality in which energy consumption is only increasing (causing increased environmental damage) and largely comes from international sources (affecting economic and national security). There are other examples, as well: In Seattle, activists pushed their mayor to commit to exceeding Kyoto carbon reduction targets; in San Francisco, voters approved a $100 million bond to support wind and solar energy projects; in Texas, the legislature mandated that up to 5 percent of energy production by 2010 would be from wind and other renewable resources; and at Tufts University, the Tufts Climate Initiative, which brings campus emissions within Kyoto limits, is being replicated by 400 different universities around the United States.[50] And, in Burlington, Vermont, grassroots efforts have succeeded in lowering electricity usage and actively worked to reduce greenhouse emissions. Equally, since 2006 marked the fiftieth anniversary of Dwight Eisenhower's "Sister Cities" initiative, a nonprofit citizen diplomacy network, similar efforts could be taken to establish "Sister Cities" climate initiatives that set examples in the "local world" for the larger world to take notice—and action.

All such bottom-up efforts, admittedly, pale against the actions, and often the failures, of larger states. One powerful example: If all the towns and cities in the United States were to mirror the energy reduction patterns of Burlington, Vermont, aggregate savings would amount roughly to 1.3 billion tons of carbon; at the same time, the lifetime emission of the newer coal plants that mainland China has brought on line amount to 25 billion tons of carbon. In other words, China's coal plants would "burn" through all of Burlington's savings (over several decades) in less than two and a half hours.[51] This brings about the recognition that bringing the "giants" (whether the United States or China) on board and in line with affecting change is crucial. As David Hawkins of the National Resources Defense Council argues, efforts to reduce pollution and carbon emissions, and use of cleanup technology, actually become industry—and state—standards over time. As industry embraces such improved technology, economic market forces drive prices down. Thus, although China is not presently building plants with carbon capture and storage (CCS), Hawkins contends that the United States should set the standard by building all new plants with CCS, creating incentives for China to follow suit—even in the absence of international treaties.[52]

We know that even if we were able to stabilize CO_2 emissions at today's levels, temperatures would still continue to rise, glacial ice sheets would continue to melt, and weather patterns would still be affected for decades to come. But levels are not going to remain at today's levels. And, although there may well be a myriad of approaches, the Union of Concerned Scientists offers a "KISS" [Keep it simple, stupid] approach to five sensible solutions to reducing CO_2 levels in this century:

1. Make Better Automobiles: better transmissions and engines, more aerodynamic designs, and stronger yet lighter material for chassis and bodies.
2. Modernize America's Electrical System: by using clean renewable energy, plant CO_2 emissions could be reduced by 60 percent, compared with government forecasts for

2020. Consumers would save a total of $440 billion—reaching $350 annually per family by 2020.

3. Continue to Increase Energy Efficiency in Homes and Businesses.
4. Protect Threatened Forests: when forests are burned or cleared, their stored carbon is released into the atmosphere. Tropical deforestation now accounts for about 20 percent of all human-caused CO_2 emissions each year.
5. Support Ingenuity.[53]

Humankind's stupidity is exceeded only by its capacity for genius. Yet, returning to the epigraph for this chapter, Abba Eban's observation that "men and nations behave wisely once they have exhausted all the other alternatives," we should remember that we have dwindling alternatives as well. Clearly, the time has come to consider those alternatives.

In the best sense, the work to address climate change and human impact combines what the World Business Council for Sustainable Development has reckoned as the organizational genius resident in the determinate authority of "GEOpolity"—scenarios stressing intergovernmental environmental agreements—and the inherent innovation of "JAZZ"—scenarios in which nongovernmental organizations, private business, and other stateless actors push forward the environmental protection agenda.[54] Equally, in terms of global cooperation, the question remains what the world would like, and what measures we should, beyond 2012—the year the Kyoto Protocol expires.

MOVING TOWARD "K2"

The United Nations' (UN's) most effective contribution to security in the twenty-first century may be to educate the people of the figurative "West"— the leaders and citizens of economically advanced and politically stable states— about where their national and global interests should lie in the future. This is a role that the UN has succeeded at in the past and can succeed again at in the future. We have one particular example in mind: the 1972 Stockholm Conference on the global environment. The conference itself was not a resounding success in the sense that the problems and issues it addressed were resolved in its aftermaths; but it was successful in formulating new, international approaches to global problems. Equally, the Stockholm Conference led to the permanent establishment of the United Nations Environment Programme (UNEP), and was the first UN organizational headquarters established in the so-called developing world. Today, UNEP serves a global environmental role, speaks out for the voiceless in promoting the disadvantages and needs of the emerging world, and (given its location in Nairobi) has been a focal point for leveraging political and civil disputes in neighboring Somalia and Sudan.

We fully acknowledge that UNEP has its share of harsh critics, most notably among them James Gustave Speth, who has suggested that UNEP (with an annual operating budget smaller than most nongovernmental organizations) has little effect in actually changing world leaders or in effecting outcomes that could bring about positive environmental change.[55] Yet even Speth singles out UNEP as "extraordinarily effective, given its size and niche in the United Nations structure."[56]

Symbolically and in reality, the Declaration of the United Nations Conference on the Human Environment[57] amounted to recognition by UN member-states that environmental issues and problems transcend borders and require national governments, local governments, civil society, industry, and the public at large to act with a common purpose. One outcome of this effort has been the decades-long effort to minimize future environmental degradation and remediate the environmental stresses of the past. UNEP, along with the UN Commission on Sustainable Development (under the UN Economic and Social Council), the various UN Conventions, and the UN University all today study and monitor the environment, mobilize civil society's engagement on environmental issues, and educate government officials at all levels about environmental science and technology.

It is true that few environmental activists will be satisfied with results to date—not only with UNEP but with all international cooperative efforts on the environment—yet environmental stresses, most especially climate change and human, will likely only accelerate in a negative direction in the future without some attempt at a massive rudder shift. In retrospect, the 1972 Stockholm Conference on the Human Environment *did* mobilize effort and spawn new approaches toward tackling environmental problems—as did the agreement most commonly known as the Kyoto Protocol. Yet although the goal of Kyoto was to require industrial nations to cut their aggregate carbon emissions by 5.2 percent below 1990 levels by 2012, even that modest amount has been seen largely as unachievable. In September 2002, British Prime Minister Tony Blair declared that "even if we could deliver on Kyoto, it will at best mean a reduction of 1 percent of global warming. But we know . . . we need a 60 percent reduction worldwide. In truth, Kyoto is not radical enough."[58] Thus, with 2012 approaching (and with it the expiration of the Kyoto Protocol), the time to not only talk—but act—for a new "K2" regime is now.

The framework and appearance of a K2 initiative need not look at all like the laborious and constricting Kyoto Protocol. Even though the United States may well opt out, again—as recent U.S. delegations have suggested might well happen[59]—the implications of climate change and human impact may by 2012 be simply too difficult to ignore or explain away. In the best of all possible worlds, it might even be possible to push toward radical attempts at solution. Ross Gelspan, for example, has argued that (given continued U.S. resistance) the Kyoto goals are dead—though the Kyoto Process is not. Thus, he calls for

a global reduction of 70 percent. Calling for a systemic set of interactive policies that could transform dependency on energies that prove environmentally
damaging, Gelspan calls for three interlinked strategies:

- change in industrial nation-state energy policies;
- creation of a "superfund" to transfer renewable energy technologies to the emerging
  world;
- subordinating, within a Kyoto-style framework, moving from (the current practice)
  on international emissions trading to a more progressively strict Fuel Efficiency
  Standards—that would rise by 5 percent yearly.[60]

One could recall the criticism made earlier in this chapter against Gelspan
and others for taking too optimistic a position, and advocating strategies that
are simply not implementable. Setting aside the feasibility of such radical strategies (though one could ask if their overall effect would still be *enough* to
prevent deep future human impact from climate change), these policy ideas
do emphasize an essential point. It is a point that the UN, of all international
organizations, should have long advocated but has simply overlooked: There is
little or no difference regarding climate change and human impact between the
developed and developing world.

Admittedly, the ingrained shadows and half-lives of memory live deeply in
our "mental maps." Consider, for example, the term "Third World," first introduced by the French demographer Alfred Sauvy in 1952 to distinguished states
that were aligned neither with the West nor with the Soviet Bloc during the
Cold War. Yet, even today, decades after the end of the Cold War, the term
"Third World" persists—as do our mental impressions of that world—and it
should have died long ago as both a distinction and a concept. Critics rightly
assail the term, as it suggests images of colonialism, the idea of the "other"; it
remains inaccurate; and it is grossly outdated. Yet its use persists, even as the idea
of the Third World has itself transformed beyond any objective classification or
distinction that marked its original usage. States such as the People's Republic of
China, the Russian Federation, and Brazil are all commonly regarded today as
Third World—even as they emerge (or reemerge) on the world stage as major
actors.

There is, of course, a more accurate depiction of the so-called developing
or Third world. It is a place where 83 percent of the world's population resides
and where 76 percent of the world's states are emerging: We should call it the
"Majority World."[61] In this way, if we accept, even partially, Ulrich Beck's conception of the "world risk society," then all misery, in the future, will be shared.
If, in the worst possible scenario, abrupt climate change occurs, borders will
become far less meaningful. There may well be only uninhabitable Diversion
Zones and more desirable Refuge Zones that distinguish the new geography
of Planet Earth.

GAIA'S REVENGE?

Michael Shellenberger and Ted Nordhaus, in their controversial monograph *The Death of Environmentalism,* as well as former vice president Al Gore, in promoting his film *An Inconvenient Truth* and in his essay that appeared in the May 2006 issue of *Vanity Fair* (titled "The Moment of Truth"), all rely on the broadly accepted (though apocryphal) notion that the Chinese word for *crisis* is a composite of the concepts of "danger" and "opportunity"—represented as either 危 枚 or 危 機.[62] And though this widely accepted cliché has entered the realm of common myth as accepted truth, there is, nonetheless, clear relevance in its correlation to climate change and human impact. Specifically, we know now that the danger of climate change is upon us; we can only hope—perhaps even pray—that the opportunity to act to avoid fatal human impacts has not passed.

Michael Crichton has argued that there are limits to "predicting the future," however, and that, at best, we may do better at simply understanding the present and managing its complexity. Following, for example, are terms that in 1900, the "environmentalist" Teddy Roosevelt did not know the meaning of:[63]

airport	jet stream	liposuction
antibiotic	shell shock	transduction
antibody	shock wave	maser
antenna	radio wave	taser
computer	microwave	laser
continental drift	tidal wave	acrylic
tectonic plates	tsunami	penicillin
zipper	IUD	Internet
radio	DVD	interferon
television	MP3	nylon
robot	MRI	rayon
video	HIV	leisure suit
virus	SUV	leotard
gene	VHS	lap dancing
proton	VAT	laparoscopy
neutron	whiplash	arthroscopy
atomic structure	wind tunnel	gene therapy
quark	carpal tunnel	bipolar
atomic bomb	fiber optics	moonwalk
nuclear energy	direct dialing	spot welding
ecosystem	dish antennas	heat-seeking
jumpsuits	gorilla	Prozac
fingerprints	corneal transplant	sunscreen
step aerobics	liver transplant	urban legends
12-step	heart transplant	rollover minutes

Yet on May 13, 1908, during the Conference on the Conservation of Natural Resources, then-President Theodore Roosevelt (presciently) pronounced words that remain equally valid in our time: "We have become great because of the lavish use of our resources . . . . But the time has come to inquire seriously what will happen when our forests are gone, when the coal, the iron, the oil and the gas are exhausted." Our "lavish" use of resources, of course, has also led to the inevitable collision between climate change and human impact as well. Toward that end, Robert F. Kennedy, Jr.—one of the fiercest environmental critics today—is far less sanguine about either the future or human willingness to change its ways: "I believe humans are hard-wired to compete, consume and ultimately destroy the planet."[64]

Crichton, in various public statements, has been quite persuasive in his argument that with so many human dilemmas, we should be working to solve and address these issues rather than banking on an uncertain future in which definite outcomes from climate change and human impact remain at worst suspect, at best uncertain. At the same time, as we have attempted consistently throughout this book to emphasize, there are advantages to challenging one's own mental maps—about both the present and the future. The Schwartz and Randall scenario on abrupt climate change, for example—as alarmist as it may well be—challenges us to reconsider how best to stem the tide of seemingly insurmountable obstacles.

The abrupt climate scenario also emphasizes—without ever directly stating—that human security is an issue of what should take on ever-increasing importance in policy choices and decisions. Human security, in short, is about protecting people.

Indeed, the human security concept centers on a concentration on the individual (rather than the state) and that individual's right to personal safety, basic freedoms, and access to sustainable prosperity. In ethical terms, human security is both a system and a systemic practice that promotes and sustains stability, security, and progressive integration of individuals within their relationships to their states, societies, and regions.

As our understanding of security concerns therefore broadens and deepens, the traditional assumption that states and governments are the sole guarantors of security will be increasingly challenged. The map of the future *will* depend on how we cope with the broader human dilemma. Addressing this dilemma requires viable sustainable development strategies and must take into account population growth, particularly in the emerging world; the rapid spread of epidemic diseases such as HIV/AIDS; the impact of climate change, including shifts in precipitation patterns and rising sea levels; water scarcity; soil erosion and desertification; and increased urbanization and the growth of "megacities" around the globe. Indeed, over the coming decades, more and more people may well be compelled by economic or environmental pressures to migrate to cities that lack the infrastructure to support rapid, concentrated population growth. (Yet we do not, today, have the means or ability to know with any certainty what numbers of "environmental refugees" we will see in the future.)

And, although it may appear tangential to our closing thoughts here on climate change and human impact, it bears worth raising the argument, as Joseph Tainter points out in his seminal work, *The Collapse of Complex Societies,* that "continued investment in complexity as a problem-solving strategy yields a declining marginal return."[65] Such logic, admittedly worthy of debate, flies in the face of Crichton and other eco-skeptics who are more interested in investing in emergency management complexity than addressing the uncertain, though increasingly threatening, future possibilities of climate change and human impact. Moreover, Under Secretary of State Paula Dobriansky's mantra-like response that "we act, we learn, we act again"[66] may leave us woefully unprepared and increasingly falling behind. As complexity increases, the learning adaptation cycle must *decrease* as well—simply to remain at the same level of knowledge. Yet, as Tainter reminds us, collapse is not a fall from heaven into primordial chaos; rather, it is a stabilizing event that occurs when complex societies can no longer sustain themselves.[67] Collapse, in other words, is a return to a more normal human condition.

All of these difficult choices cannot—and must not—be left to policy makers alone. Political leaders are, most often, driven by more pressing crisis issues in the nearer term. Political leaders do not, and often cannot, adequately focus on the future. Rather, a bottom-up driven revolution in ideas to redraw and reshape the map of future climate change and human impact requires the combined, concerted actions of powerful governments, international agencies, nongovernmental organizations, and corporations—and people themselves. "Right decisions" must focus on the long view and not just the *next* crisis; to do this wisely requires strategic attention, strategic planning, and strategic investment.

Consider, by way of conclusion then, the implications of the claims that Michael Oppenheimer[68] places alongside the second Bush administration:

It's a sad day when the president is being told by his science adviser [John H. Marburger] that climate change isn't worth avoiding. It may be possible for rich nations and people to adapt, but 90 percent of humanity doesn't have the resources to deal with climate change. It's unethical to condemn them just because the people in power don't want to act.[69]

If we have the ability to act to change such human vulnerability, one would think a truly global community would act on such a moral imperative. If not, then climate change—though it may not spell the end of humanity—may well mark the end of *our* humanity. And, if true, to be sure, Gaia will have had her revenge.

We do not know, with any form of precision, where we stand in this dilemma. Yet, to transmute one of the most famous sayings of Winston

Churchill—uttered when his nation had entered its darkest hour, and its very fate imperiled—we may have well reached "the end of the beginning" in the transition from the Anthropocene to the "Gaiacene."

NOTES

1. Excellent background notes on paleoclimatology and the "Milankovic´ Mechanism" are available from the National Climatic Data Center, Asheville, North Carolina; R. T. Pierrehumbert, *Principles of Planetary Science* (Chicago: Department of Geosciences, University of Chicago, 2006), http://geosci.uchicago.edu/~rtp1/ClimateBook/ClimateVol1.pdf; and notes from the course EESC 2100, *Climate Archives: The Climate Record of the Distant Past,* Department of Earth and Environmental Sciences, Columbia University, Spring 2006, http://eesc.columbia.edu/courses/ees/climate/lectures/cl_record.html; and a handout from the University of Michigan's Geological Sciences/Geography course 201, *Water, Climate, and Human Kind,* http://www-personal.umich.edu/~tjoa/Class/nov25_2.html;   http://geosci.uchicago.edu/~rtp1/ClimateBook/ClimateVol1.pdf.

2. See, for example, William F. Ruddiman, Department of Earth Science, University of Virginia, from a presentation titled "Orbital-Scale Climatic Synchroneity: A Greenhouse Mechanism," as part of the panel session 214, Interhemispheric Records of Paleoclimate Change: Low Latitude Influences on the High Latitudes, or the Other Way Around, in Pole-Equator-Pole Syntheses, for the Geological Society of America, November 2003. For further background on data challenging the certain astronomical validity of the "Milankovic Mechanism," see Richard A. Kerr, "Paleoclimatology: Upstart Ice Age Theory Gets Attentive But Chilly Hearing," *Science* 277 (July 11, 1997): pp. 183–84.

3. Jeffrey Kluger, "By Any Measure, Earth Is at . . . the Tipping Point," *Time,* April 3, 2006, p. 37.

4. Excerpted from the International Geosphere-Biosphere Programme (IGBP) *Newsletter41,* http://www.mpch-mainz.mpg.de/~air/anthropocene/. Crutzen, whose work has focused on ozone depletion, was influential in supporting work that led to the international treaty titled the "Montreal Protocol on Substances that Deplete the Ozone Layer." He received the Nobel Prize for chemistry in 1995.

5. Copies of the 2001 IPCC Report, which include the *Synthesis Report* and *Impacts, Adaptation and Vulnerability,* are available at: http://www.ipcc.ch/.

6. Extracted from Elizabeth Kolbert, *Field Notes from a Catastrophe: Man, Nature, and Climate Change* (New York: Bloomsbury, 2006), pp. 10–11. As Kolbert notes, Charmey was the first meteorologist (in the 1940s) to suggest that numerical weather forecasting was viable.

7. H. Sterling Burnett, "The Collapsing Scientific Cornerstones of Global Warming Theory," National Center for Policy Analysis, no. 299, June 30, 1999, http://www.ncpa.org/ba/ba299.html.

8. Jules Boykoff and Maxwell Boykoff, "Journalistic Balance as Global Warming Bias: Creating Controversy Where Science Finds Consensus," *Fairness and Accuracy in Reporting,* Fair.org. http://www.fair.org/index.php?page=1978.

9. Kolbert, *Field Notes from a Catastrophe*, pp. 54–55.

10. James Hansen, "The Global Warming Debate," January 1999, http://www.giss.nasa.gov/edu/.

11. See, for example, how Patrick Michaels of the University of Virginia, in testimony to the U.S. Congress in 1998, had eliminated two of Hansen's scenario projections and included what was the most extreme scenario—in terms of potential future temperature increase—in order to propagate his own argument that the Kyoto Protocol was "a useless appendage to an irrelevant treaty." http://www.giss.nasa.gov/edu/.

12. Marc Morano, "NASA Expert Tells Alarmists to Cool Down Climate Hype," April 24, 2006, *Cybercast News Service*. CNSNews.com.

13. Burnett, National Center for Policy Analysis, no. 299, June 30, 1999.

14. Ibid.

15. Andrew C. Revkin, "Climate Expert Says NASA Tried to Silence Him," *New York Times*, January 29, 2006, available at http://www.nytimes.com.

16. Ibid.

17. Bjørn Lomberg, *The Skeptical Environmentalist: Measuring the Real State of the World* (Cambridge, UK: Cambridge University Press, 1998), p. 318.

18. S. Fred Singer, *Hot Talk Cold Science: Global Warming's Unfinished Debate* (Oakland, CA: Independent Institute, 1998). Notably, the author of the foreword for this book is Frederick Seitz.

19. Ross Gelspan, *The Heat Is On: The Climate Crisis, The Cover-Up, the Prescription* (New York: Perseus Books, 1998); see, in particular, pp. 197–237.

20. Richard Posner, *Catastrophe: Risk and Response* (Oxford, UK: Oxford University Press, 2004), 58.

21. P. H. Liotta acknowledges that he is an expert reviewer for the forthcoming IPCC Fourth Assessment and would argue that the *Synthesis Report* produced with each IPCC publication are equally valuable to policymakers and the informed reader—one not necessarily a specialist.

22. Within the State Department, the variant on this mind-set is: "If it's been on your desk for more than ten minutes, you're now the expert."

23. Ross Gelspan, *Boiling Point: How Politicians, Big Oil and Coal, Journalists, and Activists Have Fueled the Climate Crisis—And What We Can Do to Avert Disaster* (New York: Basic Books, 2004), p. ix.

24. Mark Hertsgaard, "While Washington Slept," *Vanity Fair*, May 2006, pp. 239; 240–41.

25. Gelspan, *The Heat Is On*.

26. For one assessment of this dilemma, see Jules Boykoff and Maxwell Boykoff, "Journalistic Balance as Global Warming Bias: Creating Controversy Where Science Finds Consensus," *Fairness and Accuracy in Reporting*, Fair.org. http://www.fair.org/index.php?page=1978.

27. Bill McKibben, *The End of Nature: Tenth Anniversary Edition* (New York: Anchor Books, 1997).

28. Mark Hertsgaard, "Pentagon Warning of a New Ice Age." *Sacramento Bee*, February 22, 2004.

29. Mark Hertsgaard, "A New Ice Age?" *The Nation*, February, 2004, http://www.thenation.com/doc/20040301/hertsgaard.

30. David Stipp, "The Pentagon's Weather Nightmare," *Fortune,* February 9, 2004, pp. 100–108.

31. Edward Ortiz, "Pentagon Report Plans for Climate Catastrophe," *Providence Journal* (Rhode Island), March 3, 2004, Section B, p. 1.

32. Keay Davidson, *San Francisco Chronicle,* February 27, 2004.

33. Andrew C. Revkin, "The Sky Is Falling! Say Hollywood and, Yes, the Pentagon," *New York Times,* February 29, 2004, Section 4, p. 5.

34. Mark Townsend and Paul Harris, *The Observer,* February 22, 2004. http://observer.guardian.co.uk/international/story/0,,1153513,00.html.

35. Paul F. Stifflemire, Jr., CNSNews.com Commentary, *CyberCast News Service,* February 10, 2004. http://www.cnsnews.com/ViewCommentary.asp?Page = / Commentary/archive/200402/COM20040210e.html

36. This work was precursor to *Field Notes from a Catastrophe: Man, Nature, and Climate Change* (New York: Bloomsbury, 2006).

37. "A Planetary Problem," interview with Amy Davidson, *New Yorker,* April 25, 2005. http://www.newyorker.com/online/content/articles/050425on_onlineonly01.

38. According to Source Watch, a project of The Center for Media & Democracy, "Prior to its disbanding in early 2002, [the Global Climate Coalition] collaborated extensively with a network that included industry trade associations, 'property rights' groups affiliated with the anti-environmental Wise Use movement, and fringe groups such as Sovereignty International, which believes that global warming is a plot to enslave the world under a United Nations-led "world government." http://www.sourcewatch. org/index.php?title=Global_Climate_Coalition. Dobriansky was originally appointed Undersecretary of State for Global Affairs in 2001, and did not assume the title of Undersecretary of State for Global Affairs and Democracy until 2005.

39. Greenpeace obtained this alleged memo, reported in Chris Mooney, "Some Like It Hot," *Mother Jones,* May/June 2005, available at www.motherjones.com.

40. "GE Loses Bid to Block Global Warming Shareholder Resolution; Mutual Fund Calls On GE to Stop Advocating Global Warming Regulation," PR Web, February 1, 2006. http://www.freeenterpriseactionfund.com/release020106.htm.

41. Ibid. According to Source Watch, a project of The Center for Media & Democracy: "The Free Enterprise Action Fund is a mutual fund set up by Steven Milloy—a climate skeptic and a paid advocate for Phillip Morris, ExxonMobil and other corporations." http://www.sourcewatch.org/index.php?title=Free_Enterprise_Action_Fund.

42. Tim Appenzeller, "The End of Cheap Oil," *National Geographic,* June 2004, p. 89.

43. Elizabeth Kolbert, "Wasted Energy," *New Yorker,* April 18, 2005, p. 56.

44. William D. Nordhaus and Joseph Boyer, *Warming the World: Economic Models of Global Warming* (Cambridge, MA: Massachusetts Institute of Technology Press, 2003), pp. 130–32, 163, 178. Nordhaus, notably, is the father of Ted Nordhaus, coauthor with Michael Shellenberger of *The Death of Environmentalism: Global Warming Politics in a Post-Environmental World,* http://www.thebreakthrough.org/images/Death_of_Environmentalism.pdf.

45. Lomberg, pp. 310; 317–18.

46. http://www.safehaven.com/showarticle.cfm?id = 4677&pv = 1.

47. Posner, 44.

48. "A Planetary Problem."

49. Shellenberger and Nordhaus, 25.

50. Extracted from Gelspan, *Boiling Point,* 140.

51. Kolbert, *Field Notes from a Catastrophe: Man, Nature, and Climate Change,* pp. 177–78.

52. Ibid., 179–180.

53. Adapted from Union of Concerned Scientists Web site "Global Warming: Global Warming Solutions." http://www.ucsusa.org/global_warming/solutions/.

54. World Business Council for Sustainable Development, "Executive Summary," in *Exploring Sustainable Development: Global Scenarios: 2000–2050* (Geneva: WBCSD, 1997), 18.

55. See, for example, Speth, *Red Sky at Morning: America and the Crisis of the Global Environment* (New Haven, CT: Yale University Press, 2004).

56. Ibid., pp. 222–23. UNEP's Web site is: http://www.unep.org. Speth also acknowledges the achievements of the UN Framework on Climate Change (http://www.unfcc/int), the UN Convention on Biological Diversity (http://www.biodiv.org), the UN Convention to Combat Desertification (http://www.unccd.int/main.php), and the United Nations University (http://www.ias.undu.edu) for their thoughtful work with limited resources. Since the publication of Speth's work, the United Nations University-Institute for Environment and Human Security was founded in Bonn, Germany (http://www.ehs.unu.edu/).

57. The text of the Declaration may be accessed at the United Nations Environment Programme Web site at http://www.unep.org/Documents/Default.asp?DocumentID=97&ArticleID=1503.

58. Quoted in Gelspan, *Boiling Point,* 127.

59. Referenced at length in chapter 5; for a brief synopsis, see Juliet Eliperin, "U.S. Won't Join in Binding Climate Talks: Administration Agrees to Separate Dialogue," *Washington Post,* December 10, 2005, available at http://www.washingtonpost.com.

60. Gelspan, *Boiling Point,* 127.

61. Useful notes are available on "Third World" and "Majority World" at http://en.wikipedia.org/wiki/Third_World and http://en.wikipedia.org/wiki/Majority_World. The irony remains that Suavy's distinction in describing the "Third World" was itself a throwback to a "mental map" image of the "Third Estate of the French Revolution—where the First Estate represented nobility; the Second, the clergy; and the third, commoners."

62. For the best refutation of these symbols as representing both danger and opportunity, see the essay by Victor H. Mair, professor of Chinese language and literature at the University of Pennsylvania at http://www.pinyin.info/chinese/crisis.html.

63. Referenced with permission from Crichton's address to the National Press Club, "The Impossibility of Prediction," January 25, 2005, http://www.crichton-official.com/speeched/npc-speech.html. Transcripts of Crichton's address on "Fear, Complexity, Environmental Management: to the Washington Center on Complexity and Public Policy (November 6, 2005) and testimony before the United States Senate (September 28, 2005) are also available on the official Web site.

64. "Crimes Against Nature," Interview with Jeff Fleischer, *Mother Jones,* October 7, 2004, http://www.motherjones.com/news/qa/2004/10/09_402.html. Kennedy has argued that the increasing strength of hurricanes such as Katrina are directly linked to

global warming and has frequently asserted that U.S. government refusal to limit CO_2 output has contributed to higher levels of carbon dioxide in the atmosphere.

65. Joseph A. Tainter, *The Collapse of Complex Societies* (New Studies in Archaeology) (Cambridge, UK: Cambridge University Press, 1988), p. 120.

66. Quoted in Elizabeth Kolbert, *Field Notes from a Catastrophe: Man, Nature, and Climate Change,* pp. 149, 150.

67. Tainter, p. 198.

68. Albert G. Milbank, Professor of Geosciences and International Affairs, Woodrow Wilson School of Public and International Affairs, Princeton University.

69. Quoted in Mark Hertsgaard, "While Washington Slept," *Vanity Fair,* May 2006, p. 242. Notably, Marburger has previously surfaced in this work in reference to the "strangelet scenario" (chapter 4) and his previously expressed interest in "adaptation" rather than response (chapter 5) to climate change.

Appendix One
An Abrupt Climate Change Scenario and Its Implications for United States National Security October 2003
By Peter Schwartz and Doug Randall

This document was prepared for the U.S. Department of Defense Office of Net Assessment, by Global Business Network (GBN).

IMAGINING THE UNTHINKABLE

The purpose of this report is to imagine the unthinkable – to push the boundaries of current research on climate change so we may better understand the potential implications on United States national security.

We have interviewed leading climate change scientists, conducted additional research, and reviewed several iterations of the scenario with these experts. The scientists support this project, but caution that the scenario depicted is extreme in two fundamental ways. First, they suggest the occurrences we outline would most likely happen in a few regions, rather than on globally. Second, they say the magnitude of the event may be considerably smaller.

We have created a climate change scenario that although not the most likely, is plausible, and would challenge United States national security in ways that should be considered immediately.

Executive Summary

There is substantial evidence to indicate that significant global warming will occur during the 21st century. Because changes have been gradual so far, and are projected to be similarly gradual in the future, the effects of global warming have the potential to be manageable for most nations. Recent research, however, suggests that there is a possibility that this gradual global warming could lead to a relatively abrupt slowing of the ocean's thermohaline conveyor, which could lead to harsher winter weather conditions, sharply reduced soil moisture, and more intense winds in certain regions that currently provide a significant fraction of the world's food production. With inadequate preparation, the result could be a significant drop in the human carrying capacity of the Earth's environment.

The research suggests that once temperature rises above some threshold, adverse weather conditions could develop relatively abruptly, with persistent changes in the atmospheric circulation causing drops in some regions of 5–10 degrees Fahrenheit in a single decade. Paleoclimatic evidence suggests that altered climatic patterns could last for as much as a century, as they did when the ocean conveyor collapsed 8,200 years ago, or, at the extreme, could last as long as 1,000 years as they did during the Younger Dryas, which began about 12,700 years ago.

In this report, as an alternative to the scenarios of gradual climatic warming that are so common, we outline an abrupt climate change scenario patterned after the 100- year event that occurred about 8,200 years ago. This abrupt change scenario is characterized by the following conditions:

- Annual average temperatures drop by up to 5 degrees Fahrenheit over Asia and North America and 6 degrees Fahrenheit in northern Europe.
- Annual average temperatures increase by up to 4 degrees Fahrenheit in key areas throughout Australia, South America, and southern Africa.
- Drought persists for most of the decade in critical agricultural regions and in the water resource regions for major population centers in Europe and eastern North America.
- Winter storms and winds intensify, amplifying the impacts of the changes. Western Europe and the North Pacific experience enhanced winds.

The report explores how such an abrupt climate change scenario could potentially de-stabilize the geo-political environment, leading to skirmishes, battles, and even war due to resource constraints such as:

1. Food shortages due to decreases in net global agricultural production
2. Decreased availability and quality of fresh water in key regions due to shifted precipitation patters, causing more frequent floods and droughts
3. Disrupted access to energy supplies due to extensive sea ice and storminess

As global and local carrying capacities are reduced, tensions could mount around the world, leading to two fundamental strategies: defensive and offensive. Nations with the resources to do so may build virtual fortresses around their countries, preserving resources for themselves. Less fortunate nations especially those with ancient enmities with their neighbors, may initiate in struggles for access to food, clean water, or energy. Unlikely alliances could be formed as defense priorities shift and the goal is resources for survival rather than religion, ideology, or national honor.

This scenario poses new challenges for the United States, and suggests several steps to be taken:

- Improve predictive climate models to allow investigation of a wider range of scenarios and to anticipate how and where changes could occur
- Assemble comprehensive predictive models of the potential impacts of abrupt climate change to improve projections of how climate could influence food, water, and energy
- Create vulnerability metrics to anticipate which countries are most vulnerable to climate change and therefore, could contribute materially to an increasingly disorderly and potentially violent world.
- Identify no-regrets strategies such as enhancing capabilities for water management
- Rehearse adaptive responses
- Explore local implications
- Explore geo-engineering options that control the climate.

There are some indications today that global warming has reached the threshold where the thermohaline circulation could start to be significantly impacted. These indications include observations documenting that the North Atlantic is increasingly being freshened by melting glaciers, increased precipitation, and fresh water runoff making it substantially less salty over the past 40 years.

This report suggests that, because of the potentially dire consequences, the risk of abrupt climate change, although uncertain and quite possibly small, should be elevated beyond a scientific debate to a U.S. national security concern.

INTRODUCTION

When most people think about climate change, they imagine gradual increases in temperature and only marginal changes in other climatic conditions,

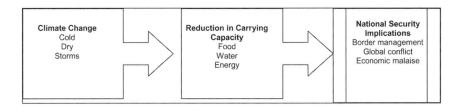

Climate Change Cold Dry Storms Reduction in Carrying Capacity Food Water Energy National Security Implications Border management Global conflict Economic malaise4

continuing indefinitely or even leveling off at some time in the future. The conventional wisdom is that modern civilization will either adapt to whatever weather conditions we face and that the pace of climate change will not overwhelm the adaptive capacity of society, or that our efforts such as those embodied in the Kyoto protocol will be sufficient to mitigate the impacts. The IPCC documents the threat of gradual climate change and its impact to food supplies and other resources of importance to humans will not be so severe as to create security threats. Optimists assert that the benefits from technological innovation will be able to outpace the negative effects of climate change.

Climatically, the gradual change view of the future assumes that agriculture will continue to thrive and growing seasons will lengthen. Northern Europe, Russia, and North America will prosper agriculturally while southern Europe, Africa, and Central and South America will suffer from increased dryness, heat, water shortages, and reduced production. Overall, global food production under many typical climate scenarios increases. This view of climate change may be a dangerous act of selfdeception, as increasingly we are facing weather related disasters—more hurricanes, monsoons, floods, and dry-spells—in regions around the world.

Weather-related events have an enormous impact on society, as they influence food supply, conditions in cities and communities, as well as access to clean water and energy. For example, a recent report by the Climate Action Network of Australia projects that climate change is likely to reduce rainfall in the rangelands, which could lead to a 15 per cent drop in grass productivity. This, in turn, could lead to reductions in the average weight of cattle by 12 per cent, significantly reducing beef supply. Under such conditions, dairy cows are projected to produce 30% less milk, and new pests are likely to spread in fruit-growing areas. Additionally, such conditions are projected to lead to 10% less water for drinking. Based on model projections of coming change conditions such as these could occur in several food producing regions around the world at the same time within the next 15–30years, challenging the notion that society's ability to adapt will make climate change manageable.

With over 400 million people living in drier, subtropical, often over-populated and economically poor regions today, climate change and its follow-on effects pose a severe risk to political, economic, and social stability. In less prosperous regions, where countries lack the resources and capabilities required to adapt quickly to more severe conditions, the problem is very likely to be exacerbated. For some countries, climate change could become such a challenge that mass emigration results as the desperate peoples seek better lives in regions such as the United States that have the resources to adaptation.

Because the prevailing scenarios of gradual global warming could cause effects like the ones described above, an increasing number of business leaders, economists, policy makers, and politicians are concerned about the projections for further change and are working to limit human influences on the climate. But, these efforts may not be sufficient or be implemented soon enough.

Rather than decades or even centuries of gradual warming, recent evidence suggests the possibility that a more dire climate scenario may actually be unfolding. This is why GBN is working with OSD to develop a plausible scenario for abrupt climate change that can be used to explore implications for food supply, health and disease, commerce and trade, and their consequences for national security.

While future weather patterns and the specific details of abrupt climate change cannot be predicted accurately or with great assurance, the actual history of climate change provides some useful guides. Our goal is merely to portray a plausible scenario, similar to one which has already occurred in human experieince, for which there is reasonable evidence so that we may further explore potential implications for United States national security.

CREATING THE SCENARIO: REVIEWING HISTORY

The Cooling Event 8,200 Years Ago

The climate change scenario outlined in this report is modeled on a century-long climate event that records from an ice core in Greenland indicate occurred 8,200 years ago. Immediately following an extended period of warming, much like the phase we appear to be in today, there was a sudden cooling . Average annual temperatures in Greenland dropped by roughly 5 degrees Fahrenheit, and temperature decreases nearly this large are likely to have occurred throughout the North Atlantic region. During the 8,200 event severe winters in Europe and some other areas caused glaciers to advance, rivers to freeze, and agricultural lands to be less productive. Scientific evidence suggests that this event was associated with, and perhaps caused by, a collapse of the ocean's conveyor following a period of gradual warming.

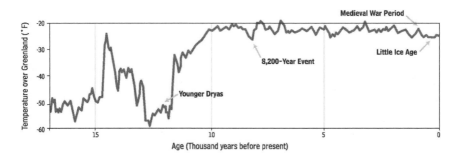

The above graphic, derived from sampling of an ice core in Greenland, shows a historical tendency for particular regions to experience periods of abrupt cooling within periods of general warming.[1]

Longer ice core and oceanic records suggest that there may have been as many as eight rapid cooling episodes in the past 730,000 years, and sharp reductions in the ocean conveyer—a phenomenon that may well be on the horizon—are a likely suspect in causing such shifts in climate.

The Younger Dryas

About 12,700 years ago, also associated with an apparent collapse of the thermohaline circulation, there was a cooling of at least 27 degrees Fahrenheit in Greenland, and substantial change throughout the North Atlantic region as well, this time lasting 1,300 years. The remarkable feature of the Younger Dryas event was that it happened in a series of decadal drops of around 5 degrees, and then the cold, dry weather persisted for over 1,000 years. While this event had an enormous effect on the ocean and land surrounding Europe (causing icebergs to be found as far south as the coast of Portugal), its impact would be more severe today—in our densely populated society. It is the more recent periods of cooling that appear to be intimately connected with changes to civilization, unrest, inhabitability of once desirable land, and even the demise of certain populations.

The Little Ice Age

Beginning in the 14th century, the North Atlantic region experienced a cooling that lasted until the mid-19th century. This cooling may have been caused by a significant slowing of the ocean conveyor, although it is more generally thought that reduced solar output and/or volcanic eruptions may have prompted the oceanic changes. This period, often referred to as the Little Ice Age, which lasted from 1300 to 1850, brought severe winters, sudden climatic shifts, and profound agricultural, economic, and political impacts to Europe.

The period was marked by persistent crop failures, famine, disease, and population migration, perhaps most dramatically felt by the Norse, also known as the Vikings, who inhabited Iceland and later Greenland. Ice formations along the coast of Greenland prevented merchants from getting their boats to Greenland and fisherman from getting fish for entire winters. As a result, farmers were forced to slaughter their poorly fed livestock—because of a lack of food both for the animals and themselves—but without fish, vegetables, and grains, there was not enough food to feed the population.

Famine, caused in part by the more severe climatic conditions, is reported to have caused tens of thousands of deaths between 1315 and 1319 alone. The general cooling also apparently drove the Vikings out of Greenland—and some say was a contributing cause for that society's demise. While climate crises like the Little Ice Age aren't solely responsible for the death of civilizations, it's undeniable that they have a large impact on society. It has been less than 175 years since 1 million people died due to the Irish Potato famine, which also was induced in part by climate change.

A CLIMATE CHANGE SCENARIO FOR THE FUTURE

The past examples of abrupt climate change suggest that it is prudent to consider an abrupt climate change scenario for the future as plausible, especially because some recent scientific findings suggest that we could be on the cusp of such an event. The future scenario that we have constructed is based on the 8,200 years before present event, which was much warmer and far briefer than the Younger Dryas, but more severe than the Little Ice Age. This scenario makes plausible assumptions about which parts of the globe are likely to be colder, drier, and windier. Although intensified research could help to refine the assumptions, there is no way to confirm the assumptions on the basis of present models.

Rather than predicting how climate change will happen, our intent is to dramatize the impact climate change could have on society if we are unprepared for it. Where we describe concrete weather conditions and implications, our aim is to further the strategic conversation rather than to accurately forecast what is likely to happen with a high degree of certainty. Even the most sophisticated models cannot predict the details of how the climate change will unfold, which regions will be impacted in which ways, and how governments and society might respond. However, there appears to be general agreement in the scientific community that an extreme case like the one depicted below is not implausible. Many scientists would regard this scenario as extreme both in how soon it develops, how large, rapid and ubiquitous the climate changes are. But history tells us that sometimes the extreme cases do occur, there is evidence that it might be and it is DOD's job to consider such scenarios.

Keep in mind that the duration of this event could be decades, centuries, or millennia and it could begin this year or many years in the future. In the climate change disruption scenario proposed here, we consider a period of gradual warming leading to 2010 and then outline the following ten years, when like in the 8,200 event, an abrupt change toward cooling in the pattern of weather conditions change is assumed to occur.

Warming Up to 2010

Following the most rapid century of warming experienced by modern civilization, the first ten years of the 21st century see an acceleration of atmospheric warming, as average temperatures worldwide rise by .5 degrees Fahrenheit per decade and by as much as 2 degrees Fahrenheit per decade in the harder hit regions. Such temperature changes would vary both by region and by season over the globe, with these finer scale variations being larger or smaller than the average change. What would be very clear is that the planet is continuing the warming trend of the late 20th century.

Most of North America, Europe, and parts of South America experience 30% more days with peak temperatures over 90 degrees Fahrenheit than they did a century ago, with far fewer days below freezing. In addition to the warming, there are erratic weather patterns: more floods, particularly in mountainous regions, and prolonged droughts in grain-producing and coastal-agricultural areas. In general, the climate shift is an economic nuisance, generally affecting local areas as storms, droughts, and hot spells impact agriculture and other climate-dependent activities. (More French doctors remain on duty in August, for example.) The weather pattern, though, is not yet severe enough or widespread enough to threaten the interconnected global society or United States national security.

Warming Feedback Loops

As temperatures rise throughout the 20th century and into the early 2000s potent positive feedback loops kick-in, accelerating the warming from .2 degrees Fahrenheit, to .4 and eventually .5 degrees Fahrenheit per year in some locations. As the surface warms, the hydrologic cycle (evaporation, precipitation, and runoff) accelerates causing temperatures to rise even higher. Water vapor, the most powerful natural greenhouse gas, traps additional heat and brings average surface air temperatures up. As evaporation increases, higher surface air temperatures cause drying in forests and grasslands, where animals graze and farmers grow grain. As trees die and burn, forests absorb less carbon dioxide, again leading to higher surface air temperatures as well as fierce and uncontrollable forest fires Further, warmer temperatures melt snow cover in mountains, open fields, high-latitude tundra areas, and permafrost throughout forests in cold-weather areas. With the ground absorbing more and reflecting less of the sun's rays, temperatures increase even higher.

By 2005 the climatic impact of the shift is felt more intensely in certain regions around the world. More severe storms and typhoons bring about higher storm surges and floods in low-lying islands such as Tarawa and Tuvalu (near New Zealand). In 2007, a particularly severe storm causes the ocean to break through levees in the Netherlands making a few key coastal cities such as The Hague unlivable. Failures of the delta island levees in the Sacramento River region in the Central Valley of California creates an inland sea and disrupts the aqueduct system transporting water from northern to southern California because salt water can no longer be kept out of the area during the dry season.

Melting along the Himalayan glaciers accelerates, causing some Tibetan people to relocate. Floating ice in the northern polar seas, which had already lost 40% of its mass from 1970 to 2003, is mostly gone during summer by 2010. As glacial ice melts, sea levels rise and as wintertime sea extent decreases, ocean waves increase in intensity, damaging coastal cities. Additionally millions of people are put at risk of flooding around the globe (roughly 4 times 2003 levels), and fisheries are disrupted as water temperature changes cause fish to migrate to new locations and habitats, increasing tensions over fishing rights.

Each of these local disasters caused by severe weather impacts surrounding areas whose natural, human, and economic resources are tapped to aid in recovery. The positive feedback loops and acceleration of the warming pattern begin to trigger responses that weren't previously imagined, as natural disasters and stormy weather occur in both developed and lesser-developed nations. Their impacts are greatest in less-resilient developing nations, which do not have the capacity built into their social, economic, and agricultural systems to absorb change.

As melting of the Greenland ice sheet exceeds the annual snowfall, and there is increasing freshwater runoff from high latitude precipitation, the freshening of waters in the North Atlantic Ocean and the seas between Greenland and Europe increases. The lower densities of these freshened waters in turn pave the way for a sharp slowing of the thermohaline circulation system.

The Period from 2010 to 2020

Thermohaline Circulation Collapse

After roughly 60 years of slow freshening, the thermohaline collapse begins in 2010, disrupting the temperate climate of Europe, which is made possible by the warm flows of the Gulf Stream (the North Atlantic arm of the global thermohaline conveyor). Ocean circulation patterns change, bringing less warm water north and causing an immediate shift in the weather in Northern Europe and eastern North America. The North Atlantic Ocean continues to be affected by fresh water coming from melting glaciers, Greenland's ice sheet, and perhaps most importantly increased rainfall and runoff. Decades of high-latitude warming cause increased precipitation and bring additional fresh water to the salty, dense water in the North, which is normally affected mainly by warmer and saltier water from the Gulf Stream. That massive current of warm water no longer reaches far into the North Atlantic. The immediate climatic effect is cooler temperatures in Europe and throughout much of the Northern Hemisphere and a dramatic drop in rainfall in many key agricultural and populated areas. However, the effects of the collapse will be felt in fits and starts, as the traditional weather patterns re-emerge only to be disrupted again—for a full decade.

The dramatic slowing of the thermohaline circulation is anticipated by some ocean researchers, but the United States is not sufficiently prepared for its effects, timing, or intensity. Computer models of the climate and ocean systems, though improved, were unable to produce sufficiently consistent and accurate information for policymakers. As weather patterns shift in the years following the collapse, it is not clear what type of weather future years will bring. While some forecasters believe the cooling and dryness is about to end, others predict a new ice age or a global drought, leaving policy makers and the public highly uncertain about the future climate and what to do, if anything. Is this merely a "blip" of little importance or a fundamental change in the Earth's climate, requiring an urgent massive human response?

Cooler, Drier, Windier Conditions for Continental Areas of the Northern Hemisphere

The Weather Report: 2010–2020

- Drought persists for the entire decade in critical agricultural regions and in the areas around major population centers in Europe and eastern North America.
- Average annual temperatures drop by up to 5 degrees Fahrenheit over Asia and North America and up to 6 degrees Fahrenheit in Europe.
- Temperatures increase by up to 4 degrees Fahrenheit in key areas throughout Australia, South America, and southern Africa.
- Winter storms and winds intensify, amplifying the impact of the changes. Western Europe and the North Pacific face enhanced westerly winds.

Each of the years from 2010–2020 sees average temperature drops throughout Northern Europe, leading to as much as a 6 degree Fahrenheit drop in ten years. Average annual rainfall in this region decreases by nearly 30%; and winds are up to 15% stronger on average. The climatic conditions are more severe in the continental interior regions of northern Asia and North America.

The effects of the drought are more devastating than the unpleasantness of temperature decreases in the agricultural and populated areas. With the persistent reduction of precipitation in these areas, lakes dry-up, river flow decreases, and fresh water supply is squeezed, overwhelming available conservation options and depleting fresh water reserves. The Mega-droughts begin in key regions in Southern China and Northern Europe around 2010 and last throughout the full decade. At the same time, areas that were relatively dry over the past few decades receive persistent years of torrential rainfall, flooding rivers, and regions that traditionally relied on dryland agriculture.

In the North Atlantic region and across northern Asia, cooling is most pronounced in the heart of winter—December, January, and February—although its effects linger through the seasons, the cooling becomes increasingly intense and less predictable. As snow accumulates in mountain regions, the cooling spreads to summertime. In addition to cooling and summertime dryness, wind pattern velocity strengthens as the atmospheric circulation becomes more zonal.

While weather patterns are disrupted during the onset of the climatic change around the globe, the effects are far more pronounced in Northern Europe for the first five years after the thermohaline circulation collapse. By the second half of this decade, the chill and harsher conditions spread deeper into Southern Europe, North America, and beyond. Northern Europe cools as a pattern of colder weather lengthens the time that sea ice is present over the northern North Atlantic Ocean, creating a further cooling influence and extending the period of wintertime surface air temperatures. Winds pick up as the atmosphere tries to deal with the stronger pole-to-equator temperature gradient. Cold air blowing across the European continent causes especially harsh conditions for agriculture. The combination of wind and dryness causes widespread dust storms and soil loss.

Signs of incremental warming appear in the southern most areas along the Atlantic Ocean, but the dryness doesn't let up. By the end of the decade, Europe's climate is more like Siberia's.

An Alternative Scenario for the Southern Hemisphere

There is considerable uncertainty about the climate dynamics of the Southern Hemisphere, mainly due to less paleoclimatic data being available than for the Northern Hemisphere. Weather patterns in key regions in the Southern Hemisphere could mimic those of the Northern Hemisphere, becoming colder, drier, and more severe as heat flows from the tropics to the Northern Hemisphere, trying to thermodynamically balance the climatic system. Alternatively, the cooling of the Northern Hemisphere may lead to increased warmth, precipitation, and storms in the south, as the heat normally transported away from equatorial regions by the ocean currents becomes trapped and as greenhouse gas warming continues to accelerate. Either way, it is not implausible that abrupt climate change will bring extreme weather conditions to many of the world's key population and growing regions at the same time—stressing global food, water, and energy supply.

The Regions: 2010 to 2020

Europe. Hit hardest by the climatic change, average annual temperatures drop by 6 degrees Fahrenheit in under a decade, with more dramatic shifts along the Northwest coast. The climate in northwestern Europe is colder, drier, and windier, making it more like Siberia. Southern Europe experiences less of a change but still suffers from sharp intermittent cooling and rapid temperature

The above graphic shows a simplified view of the weather patterns portrayed in this scenario

shifts. Reduced precipitation causes soil loss to become a problem throughout Europe, contributing to food supply shortages. Europe struggles to stem emigration out of Scandinavian and northern European nations in search of warmth as well as immigration from hard-hit countries in Africa and elsewhere.

United States. Colder, windier, and drier weather makes growing seasons shorter and less productive throughout the northeastern United States, and longer and drier in the southwest. Desert areas face increasing windstorms, while agricultural areas suffer from soil loss due to higher wind speeds and reduced soil moisture. The change toward a drier climate is especially pronounced in the southern states. Coastal areas that were at risk during the warming period remain at risk, as rising ocean levels continues along the shores. The United States turns inward, committing its resources to feeding its own population, shoring-up its borders, and managing the increasing global tension.

China. China, with its high need for food supply given its vast population, is hit hard by a decreased reliability of the monsoon rains. Occasional monsoons during the summer season are welcomed for their precipitation, but have devastating effects as they flood generally denuded land. Longer, colder winters and hotter summers caused by decreased evaporative cooling because of reduced precipitation stress already tight energy and water supplies. Widespread famine causes chaos and internal struggles as a cold and hungry China peers jealously across the Russian and western borders at energy resources.

Bangladesh. Persistent typhoons and a higher sea level create storm surges that cause significant coastal erosion, making much of Bangladesh nearly uninhabitable. Further, the rising sea level contaminates fresh water supplies inland, creating a drinking water and humanitarian crisis. Massive emigration occurs, causing tension in China and India, which are struggling to manage the crisis inside their own boundaries.

East Africa. Kenya, Tanzania, and Mozambique face slightly warmer weather, but are challenged by persistent drought. Accustomed to dry conditions, these countries were the least influenced by the changing weather conditions, but their food supply is challenged as major grain producing regions suffer.

Australia. A major food exporter, Australia struggles to supply food around the globe, as its agriculture is not severely impacted by more subtle changes in its climate. But the large uncertainties about Southern Hemisphere climate change make this benign conclusion suspect.

Impact on Natural Resources

The changing weather patterns and ocean temperatures affect agriculture, fish and wildlife, water and energy. Crop yields, affected by temperature and water stress as well as length of growing season fall by 10–25% and are less predictable as key regions shift from a warming to a cooling trend. As some agricultural pests die due to temperature changes, other species spread more readily due to the dryness and windiness—requiring alternative pesticides or treatment regiments. Commercial fishermen that typically have rights to fish in specific areas will be ill equipped for the massive migration of their prey.

With only five or six key grain-growing regions in the world (US, Australia, Argentina, Russia, China, and India), there is insufficient surplus in global food supplies to offset severe weather conditions in a few regions at the same time—let alone four or five. The world's economic interdependence make the United States increasingly vulnerable to the economic disruption created by local weather shifts in key agricultural and high population areas around the world. Catastrophic shortages of water and energy supply—both which are stressed around the globe today—cannot be quickly overcome.

IMPACT ON NATIONAL SECURITY

Human civilization began with the stabilization and warming of the Earth's climate. A colder unstable climate meant that humans could neither develop agriculture or permanent settlements. With the end of the Younger Dryas and the warming and stabilization that followed, humans could learn the rhythms of agriculture and settle in places whose climate was reliably productive. Modern civilization has never experienced weather conditions as persistently disruptive as the ones outlined in this scenario. As a result, the implications for national security outlined in this report are only hypothetical. The actual impacts would vary greatly depending on the nuances of the weather conditions, the adaptability of humanity, and decisions by policymakers.

Violence and disruption stemming from the stresses created by abrupt changes in the climate pose a different type of threat to national security than we are accustomed to today. Military confrontation may be triggered by a desperate need for natural resources such as energy, food and water rather than by conflicts over ideology, religion, or national honor. The shifting motivation for confrontation would alter which countries are most vulnerable and the existing warning signs for security threats.

There is a long-standing academic debate over the extent to which resource constraints and environmental challenges lead to inter-state conflict. While some believe they alone can lead nations to attack one another, others argue that their primary effect is to act as a trigger of conflict among countries that face pre-existing social, economic, and political tension. Regardless, it seems undeniable that severe environmental problems are likely to escalate the degree of global conflict. Co-founder and President of the Pacific Institute for Studies in Development, Environment, and Security, Peter Gleick outlines the three most fundamental challenges abrupt climate change poses for national security:

1. Food shortages due to decreases in agricultural production
2. Decreased availability and quality of fresh water due to flooding and droughts
3. Disrupted access to strategic minerals due to ice and storms

In the event of abrupt climate change, it's likely that food, water, and energy resource constraints will first be managed through economic, political, and diplomatic means such as treaties and trade embargoes. Over time though, conflicts over land and water use are likely to become more severe—and more violent. As states become increasingly desperate, the pressure for action will grow.

Decreasing Carrying Capacity

Today, carrying capacity, which is the ability for the Earth and its natural ecosystems including social, economic, and cultural systems to support the finite number of people on the planet, is being challenged around the world. According to the International Energy Agency, global demand for oil will grow by 66% in the next 30 years, but it's unclear where the supply will come from. Clean water is similarly constrained in many areas around the world. With 815 million people receiving insufficient sustenance worldwide, some would say that as a globe, we're living well above our carrying capacity, meaning there are not sufficient natural resources to sustain our behavior.

Many point to technological innovation and adaptive behavior as a means for managing the global ecosystem. Indeed it has been technological progress that has increased carrying capacity over time. Over centuries we have

Decreasing Carrying Capacity

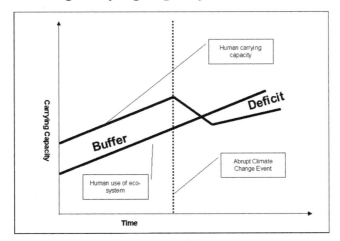

The graphic shows how abrupt climate change may cause human carrying capacity to fall below usage of the eco-system, suggesting insufficient resources leading to a contraction of the population through war, disease, and famine

learned how to produce more food, energy and access more water. But will the potential of new technologies be sufficient when a crisis like the one outlined in this scenario hits?

Abrupt climate change is likely to stretch carrying capacity well beyond its already precarious limits. And there's a natural tendency or need for carrying capacity to become realigned. As abrupt climate change lowers the world's carrying capacity aggressive wars are likely to be fought over food, water, and energy. Deaths from war as well as starvation and disease will decrease population size, which overtime, will re-balance with carrying capacity.

When you look at carrying capacity on a regional or state level it is apparent that those nations with a high carrying capacity, such as the United States and Western Europe, are likely to adapt most effectively to abrupt changes in climate, because, relative to their population size, they have more resources to call on. This may give rise to a more severe have, have-not mentality, causing resentment toward those nations with a higher carrying capacity. It may lead to finger-pointing and blame, as the wealthier nations tend to use more energy and emit more greenhouse gasses such as CO_2 into the atmosphere. Less important than the scientifically proven relationship between CO_2 emissions and climate change is the perception that impacted nations have—and the actions they take.

The Link Between Carrying Capacity and Warfare

Steven LeBlanc, Harvard archaeologist and author of a new book called Carrying Capacity, describes the relationship between carrying capacity and warfare. Drawing on abundant archaeological and ethnological data, LeBlanc argues that historically humans conducted organized warfare for a variety of reasons, including warfare over resources and the environment. Humans fight when they outstrip the carrying capacity of their natural environment. Every time there is a choice between starving and raiding, humans raid. From hunter/gatherers through agricultural tribes, chiefdoms, and early complex societies, 25% of a population's adult males die when war breaks out.

Peace occurs when carrying capacity goes up, as with the invention of agriculture, newly effective bureaucracy, remote trade and technological breakthroughs. Also a large scale die-back such as from plague can make for peaceful times—Europe after its major plagues, North American natives after European diseases decimated their populations (that's the difference between the Jamestown colony failure and Plymouth Rock success). But such peaceful periods are short-lived because population quickly rises to once again push against carrying capacity, and warfare resumes. Indeed, over the millennia most societies define themselves according to their ability to conduct war, and warrior culture becomes deeply ingrained. The most combative societies are the ones that survive.

However in the last three centuries, LeBlanc points out, advanced states have steadily lowered the body count even though individual wars and genocides have grown larger in scale. Instead of slaughtering all their enemies in the traditional way, for example, states merely kill enough to get a victory and then put the survivors to work in their newly expanded economy. States also use their own bureaucracies, advanced technology, and international rules of behavior to raise carrying capacity and bear a more careful relationship to it.

All of that progressive behavior could collapse if carrying capacities everywhere were suddenly lowered drastically by abrupt climate change. Humanity would revert to its norm of constant battles for diminishing resources, which the battles themselves would further reduce even beyond the climatic effects. Once again warfare would define human life.

Conflict Scenario Due to Climate Change

The two most likely reactions to a sudden drop in carrying capacity due to climate change are defensive and offensive.

The United States and Australia are likely to build defensive fortresses around their countries because they have the resources and reserves to achieve self-sufficiency. With diverse growing climates, wealth, technology, and abundant resources, the United States could likely survive shortened growing cycles and harsh weather conditions without catastrophic losses. Borders will be

Table A1.1 The table outlines some potential military implications of climate change

	Europe	Asia	United States
2010–2020	2012: Severe drought and cold push Scandinavian populations southward, push back from EU 2015: Conflict within the EU over food and water supply leads to skirmishes and strained diplomatic relations 2018: Russia joins EU, providing energy resources 2020: Migration from northern countries such as Holland and Germany toward Spain and Italy	2010: Border skirmishes and conflict in Bangladesh, India, and China, as mass migration occurs toward Burma 2012: Regional instability leads Japan to develop force projection capability 2015: Strategic agreement between Japan and Russia for Siberia and Sakhalin energy resources 2018: China intervenes in Kazakhstan to protect pipelines regularly disrupted by rebels and criminals.	2010: Disagreements with Canada and Mexico over water increase tension 2012: Flood of refugees to southeast U.S. and Mexico from Caribbean islands 2015: European migration to United States (mostly wealthy) 2016: Conflict with European countries over fishing rights 2018: Securing North America, U.S. forms integrated security alliance with Canada and Mexico 2020: Department of Defense manages borders and refugees from Caribbean and Europe.
2020–2030	2020: Increasing: skirmishes over water and immigration 2022: Skirmish between France and Germany over commercial access to Rhine 2025: EU nears collapse 2027: Increasing migration to Mediterranean countries such as Algeria, Morocco, Egypt, and Israel 2030: Nearly 10% of European population move to a different country	2020: Persistent conflict in South East Asia; Burma, Laos, Vietnam, India, China 2025: Internal conditions in China deteriorate dramatically leading to civil war and border wars. 2030: Tension growing between China and Japan over Russian energy ★	2020: Oil prices increase as security of supply is threatened by conflicts in Persian Gulf and Caspian 2025: Internal struggle in Saudi Arabia brings Chinese and U.S. naval forces to Gulf, in direct confrontation

strengthened around the country to hold back unwanted starving immigrants from the Caribbean islands (an especially severe problem), Mexico, and South America. Energy supply will be shored up through expensive (economically, politically, and morally) alternatives such as nuclear, renewables, hydrogen, and Middle Eastern contracts. Pesky skirmishes over fishing rights, agricultural support, and disaster relief will be commonplace. Tension between the U.S. and Mexico rise as the U.S. reneges on the 1944 treaty that guarantees water flow from the Colorado River. Relief workers will be commissioned to respond to flooding along the southern part of the east coast and much drier conditions inland. Yet, even in this continuous state of emergency the U.S. will be positioned well compared to others. The intractable problem facing the nation will be calming the mounting military tension around the world.

As famine, disease, and weather-related disasters strike due to the abrupt climate change, many countries' needs will exceed their carrying capacity. This will create a sense of desperation, which is likely to lead to offensive aggression in order to reclaim balance. Imagine eastern European countries, struggling to feed their populations with a falling supply of food, water, and energy, eyeing Russia, whose population is already in decline, for access to its grain, minerals, and energy supply. Or, picture Japan, suffering from flooding along its coastal cities and contamination of its fresh water supply, eying Russia's Sakhalin Island oil and gas reserves as an energy source to power desalination plants and energy-intensive agricultural processes. Envision Pakistan, India, and China—all armed with nuclear weapons—skirmishing at their borders over refugees, access to shared rivers, and arable land. Spanish and Portuguese fishermen might fight over fishing rights—leading to conflicts at sea. And, countries including the United States would be likely to better secure their borders. With over 200 river basins touching multiple nations, we can expect conflict over access to water for drinking, irrigation, and transportation. The Danube touches twelve nations, the Nile runs though nine, and the Amazon runs through seven.

In this scenario, we can expect alliances of convenience. The United States and Canada may become one, simplifying border controls. Or, Canada might keep its hydropower—causing energy problems in the US. North and South Korea may align to create one technically savvy and nuclear-armed entity. Europe may act as a unified block—curbing immigration problems between European nations—and allowing for protection against aggressors. Russia, with its abundant minerals, oil, and natural gas may join Europe.

In this world of warring states, nuclear arms proliferation is inevitable. As cooling drives up demand, existing hydrocarbon supplies are stretched thin. With a scarcity of energy supply—and a growing need for access—nuclear energy will become a critical source of power, and this will accelerate nuclear proliferation as countries develop enrichment and reprocessing capabilities to ensure their national security. China, India, Pakistan, Japan, South Korea, Great

Britain, France, and Germany will all have nuclear weapons capability, as will Israel, Iran, Egypt, and North Korea.

Managing the military and political tension, occasional skirmishes, and threat of war will be a challenge. Countries such as Japan, that have a great deal of social cohesion (meaning the government is able to effectively engage its population in changing behavior) are most likely to fair well. Countries whose diversity already produces conflict, such as India, South Africa and Indonesia, will have trouble maintaining order. Adaptability and access to resources will be key. Perhaps the most frustrating challenge abrupt climate change will pose is that we'll never know how far we are into the climate change scenario and how many more years—10, 100, 1000—- remain before some kind of return to warmer conditions as the thermohaline circulation starts up again. When carrying capacity drops suddenly, civilization is faced with new challenges that today seem unimaginable.

COULD THIS REALLY HAPPEN?

Ocean, land, and atmosphere scientists at some of the world's most prestigious organizations have uncovered new evidence over the past decade suggesting that the plausibility of severe and rapid climate change is higher than most of the scientific community and perhaps all of the political community is prepared for. If it occurs, this phenomenon will disrupt current gradual global warming trends, adding to climate complexity and lack of predictability. And paleoclimatic evidence suggests that such an abrupt climate change could begin in the near future.

The Woods Hole Oceanographic Institute reports that seas surrounding the North Atlantic have become less salty in the past 40 years, which in turn freshens the deep ocean in the North Atlantic. This trend could pave the way for ocean conveyor collapse or slowing and abrupt climate change.

With at least eight abrupt climate change events documented in the geological record, it seems that the questions to ask are: *When will this happen? What will the impacts be? And, how can we best prepare for it?* Rather than: *Will this really happen?*

ARE WE PREPARED FOR HISTORY
TO REPEAT ITSELF AGAIN?

There is a debate in newspapers around the globe today on the impact of human activity on climate change. Because economic prosperity is correlated with energy use and greenhouse gas emissions, it is often argued that economic progress leads to climate change. Competing evidence suggests that climate change can occur, regardless of human activity as seen in climate events that happened prior to modern society.

It's important to understand human impacts on the environment—both what's done to accelerate and decelerate (or perhaps even reverse) the tendency

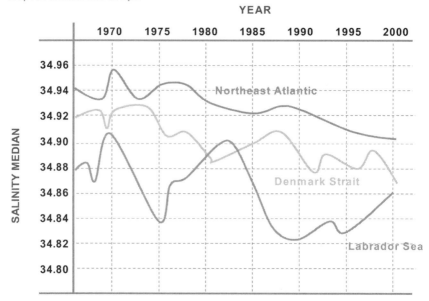

The above graphic shows early evidence that a thermohaline circulation collapse may be imminent, as the North Atlantic is increasingly being freshened by surrounding seas that have become less salty over the past 40 years.[2]

Decreasing overflow from the Nordic seas into the Atlantic Ocean through the Faroe Bank channel since 1950

Bogi Hansen*, William R. Turrell† & Svein Østerhus‡

* Faroese Fisheries Laboratory, PO Box 3051, FO-110 Tórshavn, Faroe Islands
† FRS Marine Laboratory, PO Box 101, Aberdeen AB11 9DB, UK
‡ Bjerknes Centre for Climate Research and Geophysical Institute, N-5024 Bergen, Norway

The overflow of cold, dense water from the Nordic seas, across the Greenland–Scotland ridge[1] and into the Atlantic Ocean is the main source for the deep water of the North Atlantic Ocean[2]. This flow also helps drive the inflow of warm, saline surface water into the Nordic seas[3]. The Faroe Bank channel is the deepest path across the ridge, and the deep flow through this channel accounts

Rapid freshening of the deep North Atlantic Ocean over the past four decades

Bob Dickson*, Igor Yashayaev†, Jens Meincke‡, Bill Turrell§, Stephen Dye* & Juergen Holfort‡

* Centre for Environment, Fisheries, and Aquaculture Science, Lowestoft NR33 OHT, UK
† Bedford Institute of Oceanography, Dartmouth, Nova Scotia B2Y 4A2, Canada
‡ Institut für Meereskunde, 22529 Hamburg, Germany
§ Marine Laboratory, PO Box 101, Aberdeen AB11 9DB, UK

The overflow and descent of cold, dense water from the sills of the Denmark Strait and the Faroe–Shetland channel into the North Atlantic Ocean is the principal means of ventilating the deep oceans, and is therefore a key element of the global thermohaline

The above two headlines appeared in Nature Magazine in 2001 and 2002, respectively. They suggest that the North Atlantic salinity level may lower, increasing the likelihood of a thermohaline circulation collapse toward climate change. Alternative fuels, greenhouse gas emission controls, and conservation efforts are worthwhile endeavors. In addition, we should prepare for the inevitable effects of abrupt climate change—which will likely come regardless of human activity. Here are some preliminary recommendations to prepare the United States for abrupt climate change:

1. **Improve predictive climate models.** Further research should be conducted so more confidence can be placed in predictions about climate change. There needs to be a deeper understanding of the relationship between ocean patterns and climate change. This research should focus on historical, current, and predictive forces, and aim to further our understanding of abrupt climate change, how it may happen, and how we'll know it's occurring.

2. **Assemble comprehensive predictive models of climate change impacts.** Substantial research should be done on the potential ecological, economic, social, and political impact of abrupt climate change. Sophisticated models and scenarios should be developed to anticipate possible local conditions. A system should be created to identify how climate change may impact the global distribution of social, economic, and political power. These analyses can be used to mitigate potential sources of conflict before they happen.

3. **Create vulnerability metrics.** Metrics should be created to understand a country's vulnerability to the impacts of climate change. Metrics may include climatic impact on existing agricultural, water, and mineral resources; technical capability; social cohesion and adaptability.

4. **Identify no-regrets strategies.** No-regrets strategies should be identified and implemented to ensure reliable access to food supply and water, and to ensure national security.

5. **Rehearse adaptive responses.** Adaptive response teams should be established to address and prepare for inevitable climate driven events such as massive migration, disease and epidemics, and food and water supply shortages.

6. **Explore local implications.** The first-order effects of climate change are local. While we can anticipate changes in pest prevalence and severity and changes in agricultural productivity, one has to look at very specific locations and conditions to know which pests are of concern, which crops and regions are vulnerable, and how severe impacts will be. Such studies should be undertaken, particularly in strategically important food producing regions.

7. **Explore geo-engineering options that control the climate.** Today, it is easier to warm than to cool the climate, so it might be possible to add various gases, such as hydrofluorocarbons, to the atmosphere to offset the affects of cooling. Such actions, of course, would be studied carefully, as they have the potential to exacerbate conflicts among nations.

CONCLUSION

It is quite plausible that within a decade the evidence of an imminent abrupt climate shift may become clear and reliable. It is also possible that our models will better enable us to predict the consequences. In that event the United States will need to take urgent action to prevent and mitigate some of the most significant impacts. Diplomatic action will be needed to minimize the likelihood of conflict in the most impacted areas, especially in the Caribbean and Asia. However, large population movements in this scenario are inevitable. Learning how to manage those populations, border tensions that arise and the resulting refugees will be critical. New forms of security agreements dealing specifically with energy, food and water will also be needed. In short, while the US itself will be relatively better off and with more adaptive capacity, it will find itself in a world where Europe will be struggling internally, large number so refugees washing up on its shores and Asia in serious crisis over food and water. Disruption and conflict will be endemic features of life.

NOTES

1. R. B. Alley, from *The Two Mile Time Machine,* 2000.

2. Adapted from I Yashayaev, Bedford Institute of Oceanography as seen in Abrupt Climate Change, Inevitable Surprises, National Research Council.

Appendix Two

Climate change is real

There will always be uncertainty in understanding a system as complex as the world's climate. However there is now strong evidence that significant global warming is occurring[1]. The evidence comes from direct measurements of rising surface air temperatures and subsurface ocean temperatures and from phenomena such as increases in average global sea levels, retreating glaciers, and changes to many physical and biological systems. It is likely that most of the warming in recent decades can be attributed to human activities (IPCC 2001)[2]. This warming has already led to changes in the Earth's climate.

The existence of greenhouse gases in the atmosphere is vital to life on Earth – in their absence average temperatures would be about 30 centigrade degrees lower than they are today. But human activities are now causing atmospheric concentrations of greenhouse gases – including carbon dioxide, methane, tropospheric ozone, and nitrous oxide – to rise well above pre-industrial levels. Carbon dioxide levels have increased from 280 ppm in 1750 to over 375 ppm today – higher than any previous levels that can be reliably measured (i.e. in the last 420,000 years). Increasing greenhouse gases are causing temperatures to rise; the Earth's surface warmed by approximately 0.6 centigrade degrees over the twentieth century. The Intergovernmental Panel on Climate Change (IPCC) projected that the average global surface temperatures will continue to increase to between 1.4 centigrade degrees and 5.8 centigrade degrees above 1990 levels, by 2100.

Reduce the causes of climate change

The scientific understanding of climate change is now sufficiently clear to justify nations taking prompt action. It is vital that all nations identify cost-effective steps that they can take now, to contribute to substantial and long-term reduction in net global greenhouse gas emissions.

Action taken now to reduce significantly the build-up of greenhouse gases in the atmosphere will lessen the magnitude and rate of climate change. As the United Nations Framework Convention on Climate Change (UNFCCC) recognises, a lack of full scientific certainty about some aspects of climate change is not a reason for delaying an immediate response that will, at a reasonable cost, prevent dangerous anthropogenic interference with the climate system.

As nations and economies develop over the next 25 years, world primary energy demand is estimated to increase by almost 60%. Fossil fuels, which are responsible for the majority of carbon dioxide emissions produced by human activities, provide valuable resources for many nations and are projected to provide 85% of this demand (IEA 2004)[3]. Minimising the amount of this carbon dioxide reaching the atmosphere presents a huge challenge. There are many potentially cost-effective technological options that could contribute to stabilising greenhouse gas concentrations. These are at various stages of research and development. However barriers to their broad deployment still need to be overcome.

Carbon dioxide can remain in the atmosphere for many decades. Even with possible lowered emission rates we will be experiencing the impacts of climate change throughout the 21st century and beyond. Failure to implement significant reductions in net greenhouse gas emissions now, will make the job much harder in the future.

Prepare for the consequences of climate change

Major parts of the climate system respond slowly to changes in greenhouse gas concentrations. Even if greenhouse gas emissions were stabilised instantly at today's levels, the climate would still continue to change as it adapts to the increased emission of recent decades. Further changes in climate are therefore unavoidable. Nations must prepare for them.

The projected changes in climate will have both beneficial and adverse effects at the regional level, for example on water resources, agriculture, natural ecosystems and human health. The larger and faster the changes in climate, the more likely it is that adverse effects will dominate. Increasing temperatures are likely to increase the frequency and severity of weather events such as heat waves and heavy rainfall. Increasing temperatures could lead to large-scale effects such as melting of large ice sheets (with major impacts on low-lying regions throughout the world). The IPCC estimates that the combined effects of ice melting and sea water expansion from ocean warming are projected to cause the global mean sea-level to rise by between 0.1 and 0.9 metres between 1990 and 2100. In Bangladesh alone, a 0.5 metre sea-level rise would place about 6 million people at risk from flooding.

Developing nations that lack the infrastructure or resources to respond to the impacts of climate change will be particularly affected. It is clear that many of the world's poorest people are likely to suffer the most from climate change. Long-term global efforts to create a more healthy, prosperous and sustainable world may be severely hindered by changes in the climate.

The task of devising and implementing strategies to adapt to the consequences of climate change will require worldwide collaborative inputs from a wide range of experts, including physical and natural scientists, engineers, social scientists, medical scientists, those in the humanities, business leaders and economists.

Conclusion

We urge all nations, in the line with the UNFCCC principles[4], to take prompt action to reduce the causes of climate change, adapt to its impacts and ensure that the issue is included in all relevant national and international strategies. As national science academies, we commit to working with governments to help develop and implement the national and international response to the challenge of climate change.

G8 nations have been responsible for much of the past greenhouse gas emissions. As parties to the UNFCCC, G8 nations are committed to showing leadership in addressing climate change and assisting developing nations to meet the challenges of adaptation and mitigation.

We call on world leaders, including those meeting at the Gleneagles G8 Summit in July 2005, to:

• Acknowledge that the threat of climate change is clear and increasing.

• Launch an international study[5] to explore scientifically-informed targets for atmospheric greenhouse gas concentrations, and their associated emissions scenarios, that will enable nations to avoid impacts deemed unacceptable.

• Identify cost-effective steps that can be taken now to contribute to substantial and long-term reduction in net global greenhouse gas emissions. Recognise that delayed action will increase the risk of adverse environmental effects and will likely incur a greater cost.

• Work with developing nations to build a scientific and technological capacity best suited to their circumstances, enabling them to develop innovative solutions to mitigate and adapt to the adverse effects of climate change, while explicitly recognising their legitimate development rights.

• Show leadership in developing and deploying clean energy technologies and approaches to energy efficiency, and share this knowledge with all other nations.

• Mobilise the science and technology community to enhance research and development efforts, which can better inform climate change decisions.

Notes and references

1 This statement concentrates on climate change associated with global warming. We use the UNFCCC definition of climate change, which is 'a change of climate which is attributed directly or indirectly to human activity that alters the composition of the global atmosphere and which is in addition to natural climate variability observed over comparable time periods'.

2 IPCC (2001). Third Assessment Report. We recognise the international scientific consensus of the Intergovernmental Panel on Climate Change (IPCC).

3 IEA (2004). World Energy Outlook 4. Although long-term projections of future world energy demand and supply are highly uncertain, the World Energy Outlook produced by the International Energy Agency (IEA) is a useful source of information about possible future energy scenarios.

4 With special emphasis on the first principle of the UNFCCC, which states: 'The Parties should protect the climate system for the benefit of present and future generations of humankind, on the basis of equity and in accordance with their common but differentiated responsibilities and respective capabilities. Accordingly, the developed country Parties should take the lead in combating climate change and the adverse effects thereof'.

5 Recognising and building on the IPCC's ongoing work on emission scenarios.

Academia Brasiliera de Ciências
Brazil

Royal Society of Canada,
Canada

Chinese Academy of Sciences,
China

Académié des Sciences,
France

Deutsche Akademie der Naturforscher Leopoldina, Germany

Indian National Science Academy,
India

Accademia dei Lincei,
Italy

Science Council of Japan,
Japan

Russian Academy of Sciences,
Russia

Royal Society,
United Kingdom

National Academy of Sciences,
United States of America

Selected Bibliography

In writing this book, we drew widely upon recent scholarship and several classic texts from what are often seen as three distinguishable areas of academic and professional specialization: The rapidly increasing knowledge of climate change, the expanding notion of security, and the increasingly important use of scenarios to aid decision making. Throughout the text, we have used extensive notes to lay the foundation of our thinking and build the structure of our argument, which houses these disciplines under a common roof of understanding. The following books offer insight and may be of particular interest to those who wish to continue their reading on these subjects.

CLIMATE CHANGE AND ITS INFLUENCE ON HUMANITY

Davis, Mike. *Late Victorian Holocausts: El Niño Famines and the Making of the Third World.* New York: Verso, 2001.

Diamond, Jared. *Collapse: How Societies Choose to Fail or Succeed.* New York: Viking, 2005.

Fagan, Brian M. *Floods, Famines, and Emperors: El Niño and the Fate of Civilizations.* New York: Basic Books, 1999.

———. *The Little Ice Age: How Climate Made History, 1300–1850.* New York: Basic Books, 2000.

———. *The Long Summer: How Climate Changed Civilization.* New York: Basic Books, 2004.

Flannery, Tim. *The Weather Makers: How Man is Changing the Climate and What it Means for Life on Earth*. New York: Atlantic Monthly Press, 2005.

IPCC (Intergovernmental Panel on Climate Change). *Climate Change 2001. The Scientific Basis*. Cambridge, UK: Cambridge University Press, 2001.

Kolbert, Elizabeth. *Field Notes from a Catastrophe: Man, Nature, and Climate Change*. London: Bloomsbury, 2006.

Nordhaus, William D., and Joseph Boyer. *Warming the World: Economic Models of Global Warming*. Cambridge, MA: Massachusetts Institute of Technology Press, 2003.

Posner, Richard A. *Catastrophe: Risk and Response*. Oxford, UK: Oxford University Press, 2004.

Rees, Martin. *Our Final Hour: A Scientist's Warning: How Terror, Error, and Environmental Disaster Threaten Humankind's Future in This Century—On Earth and Beyond*. New York: Basic Books, 2003.

Ruddiman, William F. *Plows, Plagues and Petroleum: How Humans Took Control of the Landscape*. Princeton, NJ: Princeton University Press, 2005.

Shellenberger, Michael, and Ted Nordhaus. *The Death of Environmentalism: Global Warming Politics in a Post-Environmental World* http://www.thebreakthrough.org/images/Death_of_Environmentalism.pdf

Wear, Spencer R. *The Discovery of Global Warming*. Cambridge, MA: Harvard University Press, 2003.

SECURITY AND ITS TWENTY-FIRST-CENTURY DIMENSIONS

Beck, Ulrich. *Risk Society: Towards a New Modernity*, trans. Mark Ritter. London: Sage Publications, 1992.

Cooper, Robert. *The Breaking of Nations: Order and Chaos in the Twenty-First Century*. New York: Atlantic Monthly Press, 2003.

Dalby, Simon. *Rethinking Geopolitics*. London: Routledge, 1998.

———. *Environmental Security*. Minneapolis: University of Minnesota Press, 2002.

Giddens, Anthony. *The Consequences of Modernity*. Stanford, CA: Stanford University Press, 1990.

Hobbes, Thomas. *The Leviathan*, ed. C. B. MacPherson. London: Penguin Books, 1985.

Homer-Dixon, Thomas. *Environment, Scarcity, and Violence*. Princeton, NJ: Princeton University Press, 1999.

Jebb, Cindy R., P. H. Liotta, Thomas Sherlock, and Ruth Margolies Beitler. *The Fight for Legitimacy: Democracy vs. Terrorism*. Greenwood, CT: Praeger, 2006.

Kaldor, Mary. *New and Old Wars: Organized Violence in a Global Era*. Palo Alto, CA: Stanford University Press, 1999.

Kuhn, Thomas S. *The Structure of Scientific Revolutions*, 3rd ed. Chicago: University of Chicago Press, 1996.

Matsumae, Tatsuro, and L. C. Chen, eds. *Common Security in Asia: New Concept of Human Security*. Tokyo: Tokai University Press, 1995.

Miskel, James F., and P. H. Liotta. *A Fevered Crescent: Security and Insecurity in the Greater Near East*. Gainesville: University Press of Florida, 2006.

Nef, Jorge. *Human Security and Mutual Vulnerability: The Global Economy of Development and Underdevelopment,* 2nd ed. Ottawa: International Development Research Centre, 1999.

Stiglitz, Joseph. *Globalization and Its Discontents.* New York: W. W. Norton, 2002.

Stoett, Peter. *Human and Global Security: An Explanation of Terms.* Toronto: University of Toronto Press, 1999.

Tehranian, Majid, ed. *Worlds Apart: Human Security and Global Governance.* London: I.B. Tauris, 1999.

Thomas, Caroline, and Peter Wilkin, eds. *Globalization, Human Security, and the African Experience.* Boulder, CO: Lynne Rienner, 1999.

THE FUTURE AND ITS UNCERTAINTIES

Bell, Wendel. *Futures Studies.* 2 vol. New Brunswick, NJ: Reaction Books, 1997.

Cornish, Edward. *Futuring: The Exploration of the Future.* Bethesda, MD: World Futures Society, 2004.

de Jouvenel, Bertrand. *The Art of Conjecture.* New York: Basic Books, 1967.

Schwartz, Peter. *The Art of the Long View.* New York: Currency Doubleday, 1991/1996.

Tainter, Jospeh A. *The Collapse of Complex Societies* (New Studies in Archaeology). Cambridge, UK: Cambridge University Press, 1988.

van der Heijden, Kees. *Scenarios: The Art of Strategic Conversation,* 2nd ed. New York: John Wiley and Sons, 2005.

Index

ABOUT THE AUTHORS

P. H. Liotta is Professor of Humanities and Executive Director of the Pell Center for International Relations and Public Policy, Salve Regina University, Newport, Rhode Island. The author of 17 books and numerous articles in fields as diverse as poetry, criticism, education, international security, intervention ethics, and foreign policy analysis, he has also published a novel about Iran. Recent work includes the coauthored *The Fight for Legitimacy: Democracy versus Terrorism* and *A Fevered Crescent: Security and Insecurity in the Greater Near East,* as well as *The Exile's Return* (published in the Macedonian language). In 2005, he became an associate of the Global Environmental Change and Human Security (GECHS) project and joined Working Group II (Impacts, Adaptation and Vulnerability) of the United Nations Intergovernmental Panel on Climate Change (IPCC).

Allan W. Shearer is an Assistant Professor of Landscape Architecture at the School of Environmental and Biological Sciences, Rutgers—The State University of New Jersey in New Brunswick. Having received his design degree and doctorate from Harvard University, his work centers on theories and methods for developing scenarios of the future. Recent research investigations have contributed to understanding potential environmental consequences of regional urbanization on and around U.S. military installations. He is a coauthor of *Alternative Futures for Changing Landscapes: The Upper San Pedro River Basin, Arizona and Sonora* and has published in several international journals including *Futures, Environment and Planning B: Planning and Design,* and *Landscape and Urban Planning.*